London Millennium Guide

By Professor S K Al Naib

This colourful London Guide, including Docklands, is a truly astonishing collection bringing to life 2000 years of the Capital's history and new millennium celebrations. It is a must for Londoners and Visitors alike.

Introduction

The Millennium Experience 2000 is to bring Britain together around the most ambitious and exciting programme of Millennium celebrations in the world. The Experience includes the Millennium Dome at Greenwich and the Millennium Festival of Great Britain. There are major exhibits designed to provide Education, Entertainment and Aspiration. The exhibitions are designed to be a shared experience for the whole nation and to make visitors feel good about Britain and its place in the next millennium.

For two years we have been working on this comprehensive guide to celebrate the new millennium and to enhance the scope of our publications on London. The aim is to help visitors enjoy London and experience the Millennium Dome Exhibitions. Chapters have been added on cultural, environmental and educational activities for the ever-increasing interests of children and parents. There are easy to read maps and extensive colourful photographs spanning from the Romans to the Millennium. I hope this guide will give pleasure and help for your trip to London.

Getting to the Dome

By tube: North Greenwich station, on the Jubilee Line, is just outside the Dome. The journey time from Waterloo is just 12 minutes.
By Docklands Light Railway (DLR): Frequent trains leave Bank underground station in the City via Tower Gateway and Canary Wharf to the Cutty Sark at Greenwich.
By boat: At peak times boats leave Westminster Pier every 15 minutes for the Dome. Other piers include Charing Cross and Tower.
By train: From Central London or the South East to Charlton Station where there is a new transit bus link to the Dome.
By car: Sorry, private cars are not allowed within two miles of the Dome. Please use public transport. Park & Ride schemes operate from the outskirts of London. Orange badge holders are allowed to park at the Dome in pre-booked spaces.
By cycle/foot: A riverside cycle/walkway links Greenwich to the Dome.

First Printing: June 1999.

Internationally Acknowledged Books by the Author

"London Millennium Guide" Ed., Ent., and Asp.	**ISBN 1 8745 36 201**
"London Dockland Guide" Heritage Panorama	**ISBN 1 8745 36 031**
"London Illustrated" History, Current & Future	**ISBN 1 8745 36 015**
"Discover London Docklands" A to Z Guide	**ISBN 1 8745 36 007**
"London Docklands" Past, Present and Future	**ISBN 1 8745 36 023**
"European Docklands" Past, Present & Future	**ISBN 0 9019 87 824**
"Dockland" Historical Survey	**ISBN 0 9089 87 800**
"Fluid Mechanics, Hydraulics and Envir. Eng."	**ISBN 1 8745 36 066**
"Applied Hydraulics, Hydrology and Envir. Eng."	**ISBN 1 8745 36 058**
"Jet Mechanics and Hydraulic Structures"	**ISBN 0 9019 87 832**
"Experimental Fluid Mechs & Hyd Modelling"	**ISBN 1 8745 36 090**

The author is Professor of Civil Engineering and Head of Department at the University of East London, England.
(Tel: 0181 590 7000/7722 ext 2478/2531, Fax: 0181 849 3423)
See information on the UEL web site at http://www.uel.ac.uk

See page 64 of this publication for ordering the books.

Printed by Lipscomb Printers, London.

CONTENTS

The Dome's central performance area.

The Millennium Dome is a feature of the skyline at night.

Canary Wharf is the jewel in London's Crown at night.

North Greenwich Station of the Jubilee Line for the Millennium Dome.

Future Canary Wharf riverside development, showing the Millennium Dome to the east.

Visiting the Millennium Dome - Visitors arriving at North Greenwich Jubilee Line station, or by coach, converge on a forecourt formed by the station and the entrance canopies. The forecourt and piazza are festive areas with kiosks and services ranging from programmes and voice guides to refreshments, banking and cloakroom facilities. Shops, cafés and restaurants are found to the north-west and east of the piazza. The Millennium Show running up to six times a day, takes place in the very centre of the Dome. One hundred and sixty performers amaze visitors with their show, taking place at heights up to 50m above the ground, with great music composed by Peter Gabriel. The show's stunning visual effects feature acrobats, trapeze artists, stilt-walkers and abseiling.

The Millennium Dome Exhibition

Education, Entertainment and Aspirations.

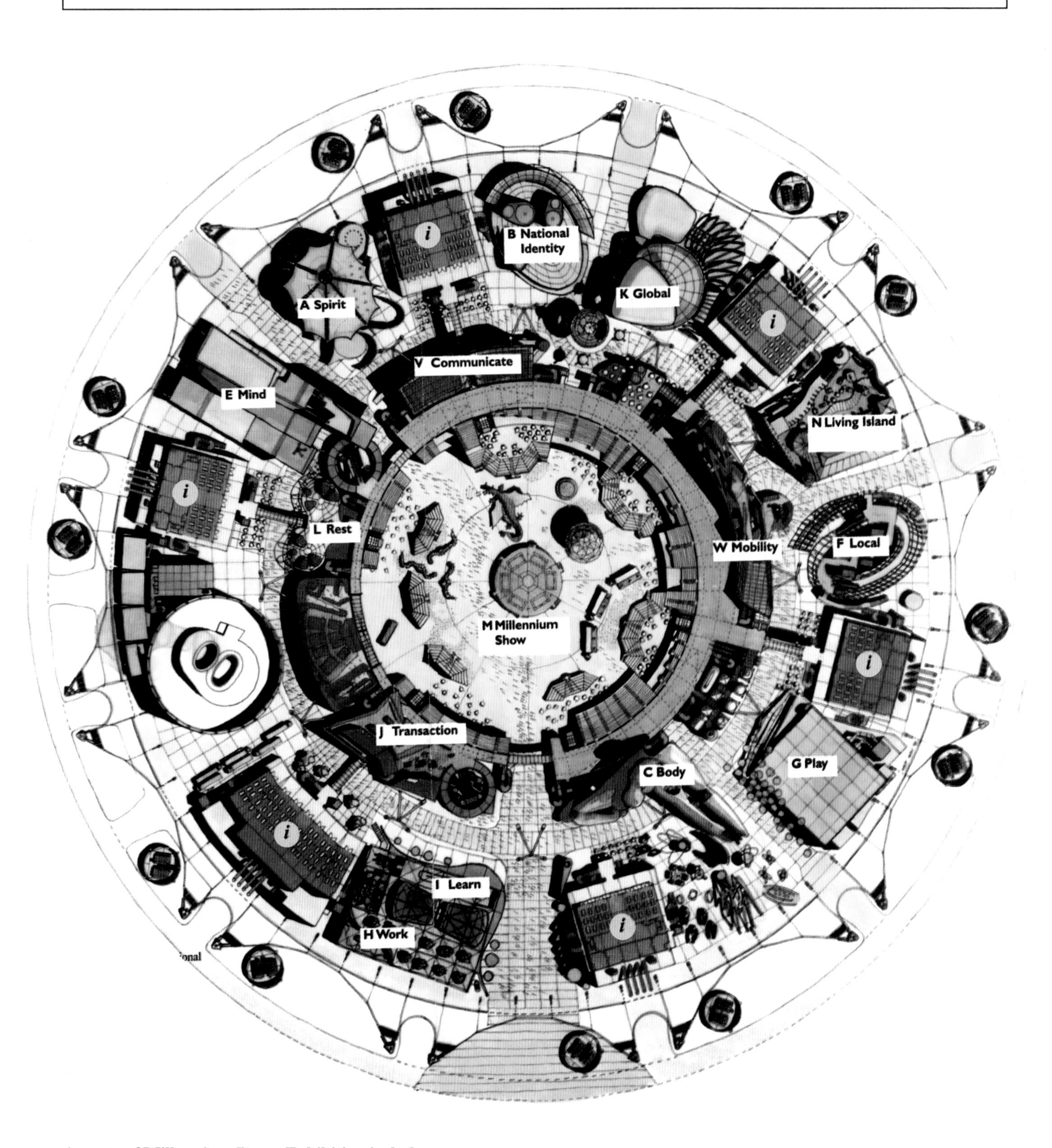

Aspects of Millennium Dome Exhibition include:

- A layout of the 14 zones in the Dome is shown in the above plan.
- GEC and British Aerospace have sponsored the £12million Mind Zone.
- Boots has put up the £12million for the Body Zone, which has a structure of two embracing figures.
- McDonalds is the sponsor of the £12million for the 'Our Town Story', a project in which towns and cities from throughout the UK stage their own events in the Dome.
- Marks & Spencer is sponsoring the Children's Promise, an initiative to persuade the UK workforce to donate their final hour's pay in 1999 to help build a better future for the children of the next Millennium. Seven of the UK's leading charities who work with children have combined to back the campaign and they share equally the proceeds for the children.

Futuristic Millennium Exhibition Zones

Britain and its place in the next millennium.

The Dome at Greenwich

Located on the Prime Meridian – longitude zero – Greenwich in London is the world's official timekeeper, marking the turning of the earth and the passing of the hours, days and years. So it is only fitting that Greenwich is the setting for the Dome, the centrepiece of the most spectacular and ambitious programme of millennium celebrations in the world. This dome construction, the largest of its kind in the world, is over one kilometre (0.6 mile) in circumference and covers over 8 hectares (20 acres). The translucent roof is 50 metres (165 ft) high at the centre and strong enough to support a jumbo jet. The Dome could contain two Wembley Stadiums or the Eiffel Tower on its side. You could even fit the Great Pyramid of Egypt inside it. The Bond movie, The World is not Enough, has been filmed at the Dome.

In fact, it contains14 vast themed zones, offering breathtaking attractions and exhibits on a scale that has never been seen before. The Dome's central performance arena, called the Millennium Show, is the setting for a spectacular live show which is repeated up to six times a day, featuring great music, stunning visual effects and a cast of up to 200 performers. The Dome opens your eyes to the astonishing possibilities that await us in the 21st century and beyond. You can begin to imagine how we might work, rest and play, what our minds and bodies might achieve, what environment we might live in. The subject matter of the individual zones is diverse, ranging from the power of finance to the Living Island, which shows how everyday choices we make affect the environment. The Spirit Zone celebrates the beliefs and traditions, while the National Identity focuses on the question of what it means to be British. Some latest technological innovations from British companies are shown. The zones are arranged as below:

A – Spirit - Take a moment to reflect in a haven of tranquillity. Explore the values that underpin our society. See how they are expressed through faith and belief.

B - National Identity - Sponsored by Marks & Spencer. A celebration of all things British – objects, places, people, attitudes, sounds, tastes – expressed through the views and perceptions of ordinary British people.

C - Body - We are now bigger, stronger, brighter and live longer. Walk through the world's largest representation of the human form and see how our amazing body works.

E – Mind - Look at the latest developments in understanding the human brain. Discover how our senses take in information and how that input is processed and stored.

F – Local - Explore real and imaginary communities from around the UK. Find out what is happening at home and abroad to change the way we live together.

G - Play - Sponsored by Sky. See the future of leisure in a fascinating playground. Experience play as a truly inspirational and fun adventure that provides unlimited possibilities for your free time.

H - Work - Sponsored by Manpower. Explore the changing world of work. See how our future career paths might evolve and change. Try special games of skill to reveal the potential that lies within you.

I - Learn - Sponsored by Tesco. Discover how vital it is for learning to remain at the heart of life, for life. Visit the classrooms of the future. See how different, challenging and rewarding the learning

J - Transaction - Learn about the impact of money on everything we do, from the personal level to the global. Watch how the money markets of the world interact and operate.

K - Global - Sponsored by BA and BAA. Take an amazing and exhilarating journey to experience some of the most remote places and extraordinary sights on earth. Understand first hand the opportunities and the responsibilities of being a citizen of the world.

L - Rest - Escape the hectic pace of life for a while. Drift off on a ride that takes you through a series of dream worlds and sensory surprises.

N - Living Island - Take a trip to the seaside to explore the relationship between the land and the ocean. See what is wonderful and what is worrying about the UK's environment today.

V - Communicate - Sponsored by BT. Learn about new approaches, technologies and ideas to help people communicate more effectively for a more fulfilling life into the 21st century.

W - Mobility - How will we get around in the future? How can technology make our journeys more enjoyable – or even reduce the need to travel at all?

The National Programme

The National Programme reaches out to every corner of the United Kingdom through a diverse series of events, activities and programmes. There are three elements to the National Programme – the Challenge, the Learning Experience and the Millennium Festival. Through these everyone in the UK has a chance to take part in events that are exciting, involving and leave a lasting legacy.

Spirit Zone

Experience a moment of peace and reflect on our deepest common beliefs.

The Spirit Zone 2000 consisting of six dramatic canopies across a formation of arches

For visitors to the Dome the central significance of the millennium celebration is that they mark the two thousandth anniversary of the birth of Christ. The Spirit Zone of the Dome in Year 2000 reflects the Christian history and nature of British society, as well as the varied multi-faithed landscape that exists in Britain today. The Spirit Zone's six dramatic canopies stretch across a formation of arches and contain an engaging and involving experience which give everyone the opportunity to explore the human relationship with faith. The Millennium Experience is an inspiring and unforgettable event for visitors.

From the earliest traces of existence humans have believed in extraordinary powers beyond their control: gods, spirits and prophets, some without form or image. The urge to spiritual association has throughout history created strong feelings and emotions which have both united and divided individuals, families, communities and nations. The Spirit Zone explores the spiritual, emotional and moral dimensions of humankind and acknowledges their importance as the animating and vital components of each person.

The Spirit Zone includes an exhibition which explores symbolically and through specific exhibits, the values that underpin our society and how they are expressed through faith. The influence of Christianity in the history of the Western world is recognised alongside the presence in our society of other religious traditions. For people of all faiths and beliefs, the millennium can encourage the shared aspiration of avoiding conflicts and building new understanding. In 1998 this Zone was called The Spirit Level and acknowledged the religious dimensions in a similar way to the final design. The Spirit Level examined faith in a giant pyramid-shaped structure formed with black glass, surrounded by a one acre garden with plants, crystals and a stream. The inspiration was drawn from monasteries, cloisters, Japanese Zen spaces and formal Muslim gardens.

The 1998 Spirit Level with a giant pyramid-shaped structure.

Body Zone

Voyage into the human machine and learn how to get the best from it.

The Body Zone comprises a male and female figures in a reclining embrace which face visitors as they enter the Dome.

The Body Zone has provoked the greatest interest among the public. The abstract figures, in a reclining embrace, face visitors as they enter the Dome. Although this is a large sculpture, it is also a building which contains a state of the art multi-media show. The spectacle begins as the visitor enters the Body at the base of the torso. The journey will take them through a series of enclosed spaces before they travel up an escalator in the body's arm to reach the main torso. The exit route then takes people through the body's legs and out into the zone's exploration area. Measuring 64 metres from the elbow to the foot and 27 metres from the ground to the top of the female's head, the structure has a surface area of 3,500 square metres. As visitors exit the sculpture, they experience thrilling sights and sounds of the remaining zones, the Our Town story performance area and a spectacular central millennium show. It is expected that up to 4000 visitors an hour will pass through the zone.

The 1998 original design for the Body Zone with a crawling baby.

The human body is the most fascinating and complex object in the world. There has been little significant evolutionary transition in its shape, structure, composition or nature since Homo Erectus became Homo Sapiens between 100,000 and 150,000 years ago, or at least not until recently. Over the last hundred years, new understanding of diet and exercise has helped transform our bodies and our lives. Today we run and jump higher, faster and longer. We are stronger, brighter, live longer and grow taller. We are also balder and more prone to cancer and allergic conditions, but have wiped out five endemic viruses in the last five years alone. We are a species that can instrumentally affect our own future and the world around us. The original design of the Body Zone also had a crawling baby.

The Body Zone is designed to amaze, with the world's biggest physical representation of the human form, a rich and intellectually accessible exploration. Visitors are taken into the world of human biology and medical science. They explore the dramatic impact of lifestyle choices on the way our bodies appear and perform, not just illness but athletic prowess, reproduction, cosmetic alterations to the body and the future of fitness.

Mind Zone

Look at the latest developments in understanding the human brain.

The giant central projection area for the Mind Zone under which visitors enter the area to explore the human brain.

Leading edge technology has been used in the design of the zones in the dome. In the Mind Zone it is not only dramatic but the gravity defying architectural design contains unique exhibits that aim to demystify the intricacies of the human brain. Visitors interact with advanced, intelligent robots in the Robot Zoo, and enjoy a specially commissioned film showing images that stimulate the very limits of our senses. British Aerospace and GEC jointly sponsor the Millennium Dome's Mind Zone. It is one of the most architecturally daring structures in the Dome and combines art and technology in its exploration of the human mind. The zone is a visually stunning piece of modern architecture. Composed of several levels, it is a fluid structure with different decks curving and interweaving with each other as they ascend towards the Dome's roof. Visitors enter the zone by passing under its giant central projection screen.

The Mind Zone celebrates the unique creativity of the human brain by exploring the nature of our senses and perceptions. This is achieved through a unique combination of traditional art and advanced technology. Technical expertise has been provided in order to ensure these attractions expand our standard visual and aural interpretations of the world around us. Specific examples include the Power of the Brain. Visitors to the Mind Zone see how brain imaging can show us which areas of our brain respond to different senses such as sight, touch, smell or sound. They see its tremendous recovery power and how it can 'rewire' itself following an accident in order to carry on functioning. Medical imaging technology is at the forefront of these developments and is used to demonstrate the amazing power of the human brain.

The most fascinating of all scientific research is into the human mind. The questions normally asked 'How do we think?' and 'When is our brain most receptive to new concepts?'. Modern technology can now show us which of those cells leap into action when a person reads, remembers or speaks. A scanner can measure blood flow in the brain and indicate where the concentration is greatest. It could be that the coming millennium will bring about a revolution in the way we think.

Our Town Story

Greenwich, along with many other towns, will also be able to tell its own Town's Story in the Dome on Our Town Story Stage. Everyday different towns from every part of the UK get the chance to tell their stories, bringing the history of towns and cities to life on the full Town Story stage. The Dome also has the performance space for 'voices of promise', a nation-wide initiative delivering the song of the millennium through the creativity of our children.

Play Zone and Entertainment Zone

Find out how leisure can re-activate your life.

The Entertainment Zone comprising two 2500-seater cinemas.

Play is in integral part of our personal, social and cultural identity. Play is not simply time spent not working. The Play Zone's bold design invites the visitor, whether adult or child, to experience the thrill of playing for themselves. The fundamental attributes of play do not change with time: the child's game, the storyteller's show, the sportsman's urge to excel. The Play Zone area is the most technologically animated area of The Dome. Huge kinetic multimedia displays rise towards the roof. Visitors join the Parallax Ride which takes them upwards on a moving pavement. Coherent images resolve from apparently formless colours and shapes: 3D objects oscillate and float within the displays, activated by other visitors ahead of you. Surrounding the ride and forming a canopy for the zone, an inflated transparent skin gently rises and falls.

The 1998 design for Serious Play.

This Zone was previously called Serious Play. This is shown as a science-fiction mountain, (see side picture), a moving pavement which transports visitors to the summit, through a swirl of 3D virtual-reality.

Situated along the side of the Dome, the Entertainment Zone consists of two 2,500-seat cinemas, one of which is convertible to a 3,000 seat performance venue. The Entertainment Zone plays host to the Film Experience, a 30-minute comedy film on the theme of time open to all visitors during the normal opening hours of the Dome. The Entertainment Zone also provides London with a new performance stage hosting a variety of evening events including concerts, conferences and one-off shows. It is to be used for introductory talks for the estimated 5000 children who visit the Dome on school trips each day.

The singers of today would have remained as groups from the back streets had Thomas Edison not demonstrated his gramophone apparatus at the Paris World Exhibition of 100 years ago. The old musical wax roller made the dream of all performers come true, they sing and millions of people listen.

The Local Zone (see page 5).

Work Zone and Learn Zone

Match your hidden skills to the new world of work and open your mind to life-long learning.

The Learn Zone for life-long learning.

We have previously approached work as involving a job for life, fuelled from a single 'tank' of education acquired during our youth. The future requires a more flexible approach to learning and work. Education must 'refuel' and re-skill us throughout our lives. The close inter-relationship of learning with the changing world of work is reflected in this design, which interweaves two zones on two levels, each offering an exploration of our future working and learning lives.

The Learn Zone tackles the topic of how we learn to learn. This zone explores the ways in which 21st century children can start their lives with a desire to go on learning. Visitors see classrooms of the future as they may develop to support learning by children and adults alike, see and talk to children in their schools world-wide and ride the Learn Zone.

In this zone we examine how we are moving away from a job for life, how people can expect more changes of employer and more changes in what they do. As employment opportunities change, so we will have to change ourselves, learning to be more flexible and acquiring skills relevant to the future. The design ranges from the animatronic orchestra of work, through the 20/20 gallery of young people's views of what the future holds, to an interactive game which brings the theme of flexibility to life.

The 1998 design of the Learning Curve showed classrooms of the future, while Licensed to Skill had an interactive game on the theme of adapting to a world where jobs are no longer for life, (see picture on the right). Up to 400 visitors at a time could use virtual reality headsets to experience how people would work.

Giant computer screens and game skills abound in the Work Zone. The idea of the computer first occurred to a Professor at Cambridge University in 1835. Charles Babbage designed the world's first mechanical computer, but the technology of the time could not advance his ideas. During the 20th century the tiny piece of engineering called the 'chip' has moved the frontiers of the field. Today and for the future these tiny elements made from pure silicon are at work in millions of computers and countless CD players, medical equipment, television and satellites all over the world.

The 1998 design for the Learning Curve and Licensed to Skill.

Rest Zone

Dream, imagine and return refreshed.

Dreamscape, the 1998 original design, presented a restful landscape of smooth pebble shapes strewn across the floor of the Dome.

Twentieth century life values action at the expense of reflection and calm. We expect to be in motion and in bloom all the time. Is rest a state of stagnation or a stage of healing, recovery, preparation and regeneration? The more we ask of ourselves, the more stress we endure and the less time we allow ourselves, the more we need to rest. We must learn new ways to rest, learn to take time to make a difference to ourselves. This zone seeks to relieve the pressure always to 'do something'. It emphasises the values of relaxation and contemplation, of letting imaginations run free. This is the place in the Dome which offers the positive benefits of rest and tranquillity, a calming 'floater-coaster'.

Previously this Zone was called Dreamscape which had a restful pebble-shaped landscape.

The Zone involves 16 people at a time climbing aboard giant beds which take a 'floater-coaster' ride on a river of dreams. The ride floats out of a bedroom window, through a flock of sheep, into clouds, over a London scene and into a seascape and serial dream sequence – each with its own sounds and smells! The dream environments intend to surprise, excite and entertain, setting minds free in a way that only dreams can achieve. Visitors get woken at the end by a giant alarm clock. This Dali-esque dream environment may offer relief from the pace and variety of a day out in The Dome.

Cinema, theatre, concert halls and television are venues for relaxation. Edward Maybridge,the English photographer, was the first person to have the idea of capturing movements in a sequence of photographs. In 1887 he demonstrated a series of photographs as a film and became practically the inventor of film projectors. The era of big screen had begun. The prototype of the television receiver met with little interest back in 1929, with a screen about the size of a postage stamp. It was not until the first electronic television was presented at the 1933 World Exhibition in Chicago that the screen was made big enough for the whole family. The results are all only too well known.

Simon Armitage is the Poet in Residence for the Millennium Experience. His task was to write the 1000 line poem to celebrate the new millennium against the Dome's fourteen zones. It is a platform to introduce restful poetry to people of all generations.

Living Island

Choose how to protect your environment day by day.

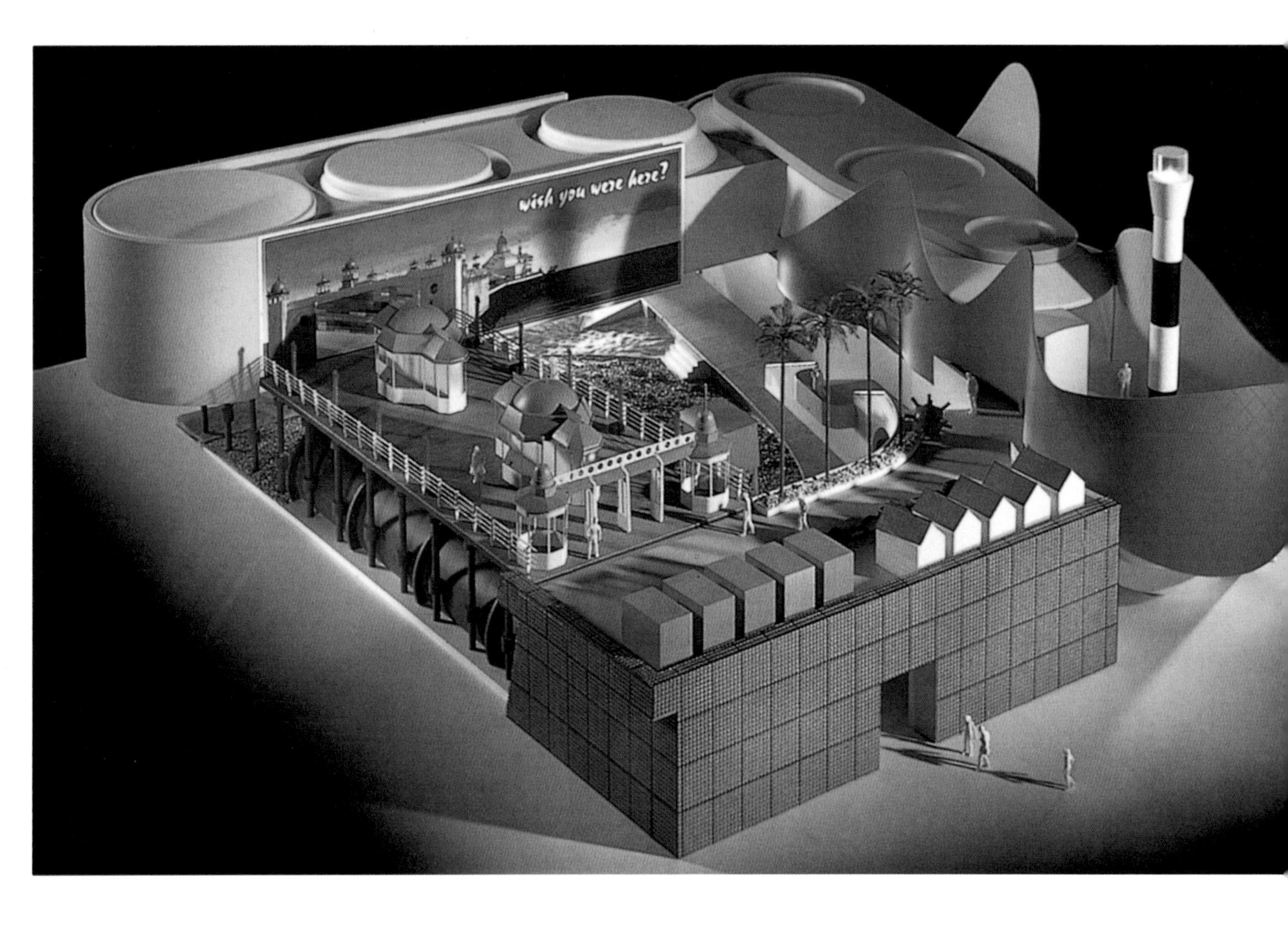

The 1998 design of Living Island represented a typical British seaside resort.

Other zones in the Dome astound and inspire, ranging from Transaction – the power of finance – to Living Island, which shows how everyday choices we make affect our environment. While the National Identity Zone asks what it means to be British in the next millennium and what you think is the best of British.

Living Island is where visitors explore the difference they can make to our environment through everyday choices. The setting will be unfamiliar to many of today's tourists: a typical British seaside resort, complete with beach fish-and-chip stall, deckchairs, sun and sea. The Living Island Zone has a giant postcard reading 'Wish You Were Here?' as the backdrop to a typical 1950s British seaside resort scene.

We move beyond the traditional view of Britain as a green and pleasant land and tour the edges of the British Isles, to explore our relationship with our environment. We all have an interest in sustaining the world's resources, yet we are reluctant to adopt new ways of living which reduce our freedom and quality of life. This zone considers the choices open to us, showing how even small changes in lifestyle, multiplied collectively, can make a large impact on the future.

The original design of 1998 took visitors on a journey to a typical British seaside resort. In a funfair and pleasure beach setting, visitors try out games, contests, rides, trials of skill and strength. The tone is intriguing and positive for adults, and hands-on fun for the young.

Greenwich Peninsular is being developed extensively for housing south of the Dome. The development plans are to create housing and a central business district on land being used by the New Millennium Experience

The future of London and Britain has many different faces. Tomorrow's technology should no longer be allowed to follow the old road of bigger and faster. Today's progress means that technology has to be properly managed and more compatible with the environment and more sustainable. The principle which will secure our future over the next century is already known. This is the concept of balanced sustainable development which can be applied to the environment, mobility and power supplies as an agenda of 21st century for the world community.

Countdown to the New Millennium

Millennium Dome

1993

April: US TV networks discussed with Greenwich Councillors the millennium prospects when they tried to book locations for world wide eve of millennium broadcast.
November: London Borough of Greenwich commissioned Price Waterhouse to produce feasibility study of millennium exhibition at Greenwich.

1994

February: Price Waterhouse reports got enthusiastic welcome from Greenwich councillors and Government. It estimated 10-15 million people would visit the millennium exhibition.

June: British Gas, owners of the exhibition site, agreed to pay towards the planned Jubilee Line station at North Greenwich, the worlds biggest underground tube station. Millennium Commission immediately confirmed there would be an exhibition, but didn't specify where.

December: London Borough of Greenwich commissioned Victor Hausner to study the economic benefits of a millennium exhibition at Greenwich.

1995

January: British Gas agreed to set aside 50 acres for the millennium exhibition in the Greenwich peninsular.

April: The Hausner Report estimated £500million capital investment and 10,000 permanent jobs.

June 1: Millennium Commission announced guidelines for site bids, aiming to announce six possible locations by the end of 1995.

June 14: Sir Bob Scott was appointed to lead the Greenwich bid.

June 28: Greenwich bid submitted, backed by Greenwich Council, Association of London Government, London Tourist Board and the Thames Gateway Boroughs.

July 14: Site shortlist was announced as Derby, NEC Birmingham, Bromley by Bow and Greenwich. NEC hyped up media storm by claiming the event at Greenwich would cause major traffic problems.

October 9: The press reported that Michael Heseltine was backing Greenwich.

October 23: The Sunday Times claimed to have uncovered an operation favouring the reclamation of contaminated land at Greenwich.

October 30: Greenwich Council launched its new transport strategy which majored on the use of the River Thames to ship in Greenwich celebration visitors.

November 30: Heritage Minister announced support for Greenwich which is to be granted World Heritage site status.

December 5: Millennium Commission short-listed operators as MAI, M2000, Granada and Imagination. The first two merge, Granada pulled out and Imagination became the front runner.

1996

January 18: Imagination - involved initially with the Birmingham bid - invited to develop proposals for an exhibition at Greenwich. NEC, Derby, and Bromley by Bow bids were rejected. Financial support campaign for Greenwich produced little other than a pledge from British Airways.

February 23: Press reports indicated the Government had chosen Greenwich for the nations celebration.

February 28: Heritage Secretary, Virginia Bottomley announced the Millennium Commission preferred site was Greenwich which would produce a substantial legacy for London and the UK.

June 18: Millennium Commission confirmed Greenwich and announced a new company, Millennium Central Limited (MCL) with BA's Robert Ayling as chairman.

July 4: The Docklands Forum and Greenwich Waterfront Community Forum staged a major millennium conference with 20 delegates with wide cross sectoral views.

August: Chris Peckham appointed as acting chief executive of MCL. Work started on de-contamination of Greenwich Peninsula site as part of a £15million programme.

October: Submission of planning application to London Borough of Greenwich DLR. £168million Greenwich and Lewisham extension was given the go ahead with Cutty Sark station included. DLR Monitoring Group protested at demolition of historic buildings in Greenwich town centre to make way for Cutty Sark station.

November: DLR Monitoring Group and English Heritage gained enquiry in council's plans to demolish historic buildings.

December: Millennium Central Limited business plan showed ticket sales to 13.5 million visitors paying an average of £20 each producing revenue of £270million. Millennium Commission confirmed grant of £200million to Greenwich Exhibition project.

1997

January 9 : Labour and Tory Chiefs agreed cost and capping for the millennium exhibition.

January 12: Millennium Commission chief executive, Jenny Page became chief executive of operating company Millennium Central

January 28: Planning application for dome approved.

January 31: Revised budgets revealed to Millennia showed £570million costs and £576million income with more private funding needed.

February: Millennium Dome contractors appointed. Off site construction began.

April 6: National Maritime Museum's launch of the "1000 Days Countdown to the Millennium".

May: Labour Party won the General Election.

September: On site construction begins.

December: The £40million regional millennium programme from Imagination began.

1998

January: Deadline for all private sector funding for Greenwich Exhibition to be in place.

April: The organisers of the Millennium Exhibition issued statements aiming to silence their critics and win greater public support after unveiling some of the exhibits and attractions that would fill the dome. The Prime Minister, Tony Blair, made a declaration of support for the project, vowing it would be the envy of the world, while announcing a number of new business sponsors.

August: Improvements start on the A102 motorway link to Greenwich and the millennium site.

September: Beginning of Millennium Dome fit out.

1999

January: New descriptions for the millennium exhibitions inside the dome were disclosed.

May: Installation of Dome Central area.

June: Millennium Pier completed.

July: First testing of the millennium dome and facilities. Tickets go on sale.

November Jubilee line Extension opening.

November: Completion of Greenwich Millennium Exhibition.

December 31: Greenwich welcoming the 21st Century. Grand Opening Ceremony.

2000

January 1: The biggest "21st" Century party continuing into the new century.

Opening of Millennium Wheel.

February: Opening of Dockland Light Railway Greenwich and Lewisham extension.

April: Opening of Millennium Footbridge.

December 31: The biggest party Britain has ever seen welcomes the new Millennium.

2001

January 1: The biggest celebration continuing into the New Millennium.

Millennium Wheel

The Millennium Wheel standing high above the Thames, is the largest observation wheel in the world, with an overall diameter of 135m to the outside of the passenger capsules. The 32 capsules, each carrying 25 people, offers a bird's eye view of London from the South Bank during the half-hour rotation, (see page 39).

Millennium Foot Bridge

Erected on the site of the 12th century London Bridge, that stood lined with houses and shops for more than 600 years, London's new river crossing provides a pedestrian route across the Thames, (see page 39). It connects the Globe Theatre and the new Tate Gallery of Modern Art on the South bank to St Paul's Cathedral in the City.

It is estimated that about four million people a year will use the walkway after its opening in April 2000. Made of stainless steel and aluminium the suspension bridge promises to offer new views of London, linking St Pauls Cathedral and the New Tate Gallery in the old Bankside Power Station, much has been made of its ability to revitalise this area of the Capital.

The impact that it has in terms of regeneration not only applies to the historic city but also to Southwark, where new activities are taking place bringing the river to life. The bridge is the first new crossing of the Thames for more than a century. The design team includes the eminent sculptor, Sir Antony Carro, whose prominent sculpture draws pedestrians to the bridge.

New Millennium Celebrations

Millennium Experience

The Millennium Celebration of 2000 is in the traditions of the Festival of Britain in 1951 and the Great Exhibition of 1851 both held in London. The Festival of Britain attracted over 8million visitors in around six months and transformed a bomb-damaged site into the South Bank Centre and Royal Festival Hall of today. The legacy of the Great Exhibition 1851 included the building of the Albert Hall, Imperial College and museums in South Kensington.

The Millennium Experience is funded through the Millennium Commission, the independent body set up to celebrate the Millennium using National Lottery funding. The purpose of the commission was to assist communities to mark the closing of the second millennium and to celebrate the start of the third. The Commission used money raised by the National Lottery to encourage projects throughout the United Kingdom. These projects enjoyed public support and were considered to be of lasting monuments to the achievements and aspirations of the British people. They were also to support a programme of awards to individuals and to enable the nation to enjoy a festival in the year 2000. The Millennium Commission decided in February 1996 that a National Millennium Exhibition would take place at Greenwich throughout the year 2000. It agreed to provide £200million to support the Experience. In January 1997 the then Conservative Government agreed with the Labour opposition to extend the life of the Millennium Commission by one year and to underwrite the cost of the Millennium Exhibition. The Millennium Central Company was established in February 1997 as a publicly owned company responsible for managing the construction and operation of the Millennium Exhibition. In June 1997 the new Labour Prime Minister, Tony Blair, visited the Greenwich site and confirmed its backing for the exhibition. It was announced that the project would be renamed the Millennium Experience and the company would become the Millennium Experience Company Limited.

The same is true in our cultural life, in sport, and in our national heritage, whether man-made or the work of nature. 2000 is, after all, the Arts Council's Year of the Artist. The period from 1995 to 2000 has stood out as a period not only of cultural renaissance, but also of renewal of our capital stock to compare with that created by the flow of imperial commerce in the 19th century.

Key themes put forward for the Millennium Commission's support included environmental conservation and restoration. The Millennium Forest in Scotland, the Earth Centre in Doncaster, the Millennium Seed Bank and the Heritage Fund programme for urban parks were good examples. Schoolchildren were asked to send their five wishes for the new Millennium where they attached great importance to global conservation. Time and again polls tell us that this is the issue which concerns young people more than any other. Environmental concerns are in any case particularly appropriate for the new Millennium, for we are both looking back to our inheritance and looking forward to the world which will be inherited by our children and grandchildren.

Peoples Celebrations

People were considered to be at the centre of the Millennium celebration. The task went beyond the creation of monuments for their own sake, or simply for the purpose of celebration. It was to ensure that people are participants rather than just spectators. The opportunity to renew the willingness on the part of people from every background was to volunteer their time, energy and expertise. People were giving and gaining, through community involvement - not only improving their communities and the country at large, but also increasing their own sense of worth.

Businesses took their corporate responsibilities to the community ever more seriously. A spirit of social entrepreneurship developed which inspired admirable projects such as the work Elaine Applebee is doing with communities in Bradford, or Andrew Mawson's work in Bromley-by-Bow.

National Lottery Funding

The National Lottery Charities Board round of grants was committed to funding volunteering, training and reconstruction. The Millennium Commission earmarked £200million for its Millennium Awards scheme, under which bursaries were made available to individuals to enable them to achieve personal goals whilst bringing a clear benefit to their communities. Grants were administered through partner bodies such as charities and other philanthropic organisations. For example, £2.7million went to the Prince's Trust to benefit 2,500 disadvantaged youngsters around the UK. Community Service Volunteers, Operation Raleigh, Mind and Help the Aged were also early partners in the initiative. This strand of the Millennium Commission's work was only just beginning, and an interest-bearing fund will support the Awards well into the 21st century, thus bringing the benefits to hundreds of thousands of people.

Millennium Dome Exhibition

Close by the magnificent Royal Naval College, the Dome reflects the best in modern British design and innovation. It is the largest structure of its type in the world, twice the area of Wembley Stadium and taller than Nelson's Column. It provides an enormous, light and airy internal space to house the Exhibition, an event of international significance and a worthy successor to the exhibitions of 1851 and 1951. It provides a unique experience for the visitor, exiting, enlightening and enjoyable. The Exhibition is to be a national asset, showing the world the breadth and quality of the British people's vision and abilities.

The entire nation is reflected in the Exhibition. In addition to the national programme relating to the Exhibition, the Millennium Commission has awarded £20million towards local celebrations around the country, and invited the public to submit suggestions for events. The consultation exercise ended in 1996 and was followed by a bidding process.

Some themes to which we can all commit ourselves - regeneration and renewal - physical, personal and social. The role of Government is to facilitate and co-ordinate discussions to draw out a sense of common purpose. The idea was to help every community in Britain to feel that the Millennium has a relevance to them.

Co-ordination of Celebrations

The Churches, the Royal Household, Whitehall departments, local authorities, the Commission and representatives of the media were joined in a Millennium Co-ordinating Group. It, in turn, gave birth to a number of sub-groups dealing with specific areas of concern. One group met at Lambeth Palace. Another example was the London sub-group. It involved key players including London First, the Government Office for London, the Association of London Government and the London Millennium Festival organisers.

Throughout the Millennial year local celebrations take place throughout the country, without Government or Commission involvement. This is particularly true of New Year's Eve 1999. A number of British local authorities are joining in the Beacon Millennium, a project which among other things involves a world-wide network of bonfires being lit on New Year's Eve 1999. Major venues in London and elsewhere had long been booked up for events. Scottish Hogmany is always a memorable affair. It is a bumper year for manufacturers of fireworks and purveyors of champagne. We know that the Exhibition has a serious educational content.

The new Millennium provides an opportunity for people of all ages and backgrounds to come together to meet their neighbours, perhaps for the first time. Street parties and fireworks can provide a one-off chance to do this. The traditional meaning of the word Millennium is the thousand-year reign of the saints. Traditionally, this would be preceded by the appearance of the Anti-Christ and an apocalyptic battle between Good and Evil.

Ideas and Inventions

On 11th October 1851 the organisers of the first world exhibition counted over six million visitors which proved that the idea has been turned into an event which acted like a magnet for inventors, manufacturers and public alike. A succession of world exhibitions over the years has enabled generations of people to experience the very best ideas of the time. From the weaving loom to the refrigerator, from the lift to the car, from supersonic aircraft to Venus probes. Much has changed in these intervening years; new ways have to be found to give our world a chance for the future. The Millennium Exhibition shows where the opportunities lie; in the creativity, inventive spirit and fantasy of the people of this planet.

It is thought that the first idea of cultivating wheat from grasses came about around 10,000 years ago. The small grains of the stone age provided food for our race which thrived and expanded over many millenniums. During the year 2000 bread and oil from the wheat forms part of the diet for several billions of people and still too few have full stomachs.

Map of UK Millennium Projects

Since 1994 the Millennium Commission had been putting together the infrastructure for the Millennium celebrations. By December 1995 it had committed just under £770million to 94 capital projects at over two thousand sites around the country, (see map on page 15), from great projects such as the new National Stadium for Wales, the Lowry Centre in Salford, Hampden Park Stadium in Glasgow and the Ballymena Town Park in County Antrim, to nearly 400 new village halls and community centres across Britain. Vital new amenities were being built from Orkney to the Isle of Wight. These are lasting monuments, physical reminders of the Millennium which has brought fresh life to villages and cities for decades to come. Our Millennial challenge, however, was to build communities and not just buildings.

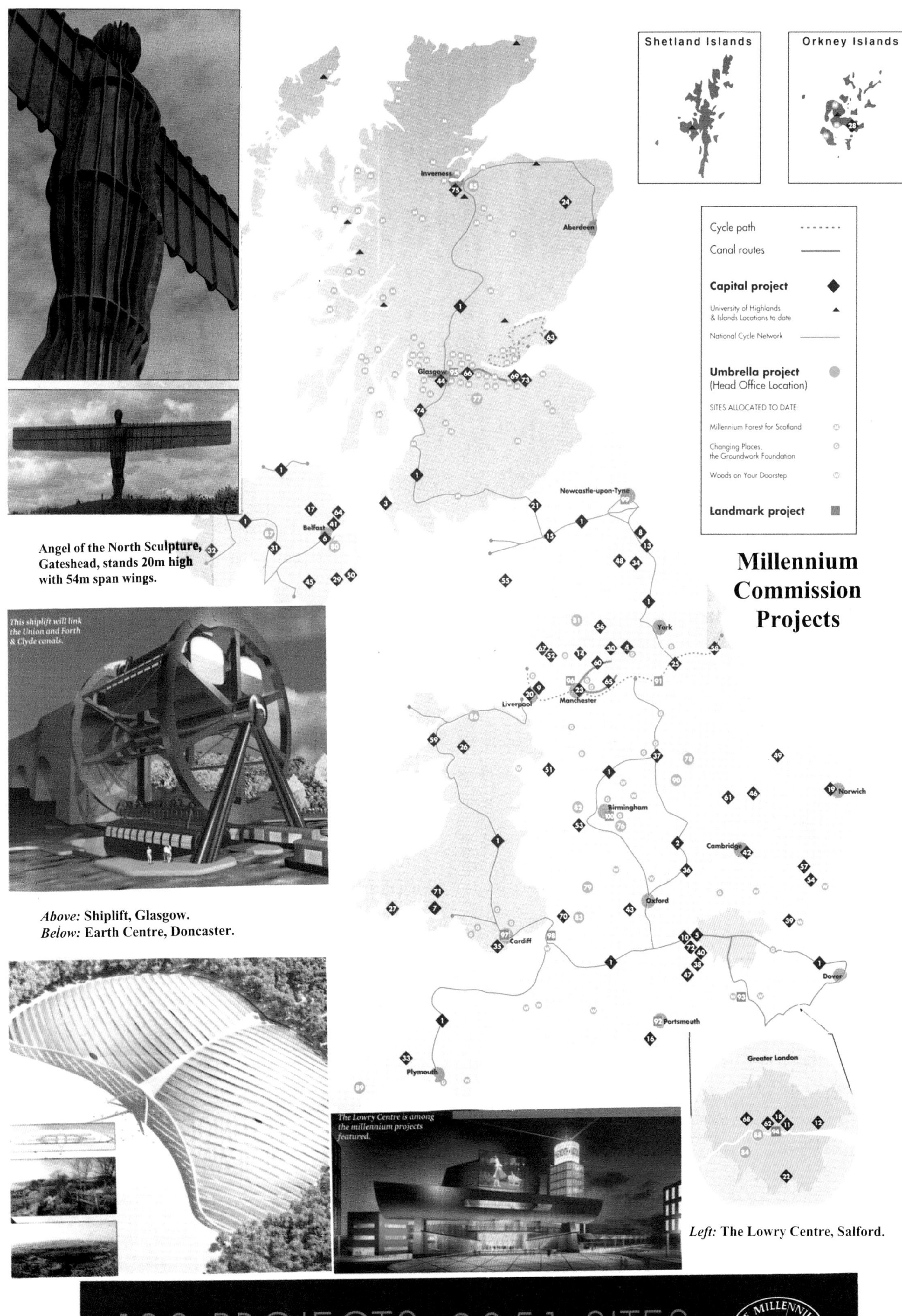

Millennium Commission Projects

Angel of the North Sculpture, Gateshead, stands 20m high with 54m span wings.

Above: **Shiplift, Glasgow.**
Below: **Earth Centre, Doncaster.**

Left: **The Lowry Centre, Salford.**

Building the Millennium Dome

Staging the Exhibition

By 1995 four sites were being considered to host the Millennium Exhibition which half the population of the UK was expected to visit in 2000. Bookmaker William Hill made Birmingham and Greenwich the favourites, but Derby and Stratford were both optimistic. The decision on where the national fair would be and who would run it rested with the Millennium Commission. Funding for the Commission came from the National Lottery - millennium schemes shared the proceeds with the arts, sports, heritage and charities. As well as the celebrations in 2000, grants were made to back capital projects.

Decision-making for the exhibition was a rather complex business, with Birmingham, Greenwich, Derby and Stratford competing for the location, and potential operators suggesting what the fair should entail. It was presumed that it would be something more serious than a fun park but far more entertaining than a trade fair. Two of the initial four potential operators remained in the running, after Granada dropped out and two others joined forces. The Birmingham National Exhibition Centre had the edge in practical terms, or so its promoters claimed. The road and rail links were already in place and over the years have demonstrated their ability to bring a Motor Show sized crowd of 125,000 a day from all around the country to the NEC halls.

Because of the area's unique connection with world time, the third millennium would effectively begin at Greenwich. It was believed that Greenwich is both the most exciting and romantic choice for the country. The Greenwich site lies just downriver of the historic Greenwich town centre and a stone's throw from Canary Wharf. The majority of the peninsula is being upgraded for the 21st century.

London Underground was extending the Jubilee Line into east London. The biggest station on the new line is the North Greenwich which allows direct access into central London. Greenwich also has an overground rail link to central London and by 2000 would have a Docklands Light Railway station. London City Airport, which serves both domestic and European destinations, is only just across the water.

Finally, after a national competition with over fifty suggested locations which boiled down to three or four locations at the end, in February 1996 the Millennium Commission selected Greenwich to be the site for the national Millennium exhibition. It was subsequently renamed the Millennium Experience.

Building the Dome

The design of the Millennium Exhibition Dome was based on two structural traditions: the Dome and the Tent. By combining these two traditions the Dome's architects were able to exploit both sentiment and sound building practice. The Dome is the strongest shape available, while the tent enables a huge area to be covered at low cost and without great difficulties of construction. For the German pavilion at Montreal Expo in 1967 a tent-like dome was constructed, the same design was used for the elegant Olympic stadium in Munich which covered 7 hectares (18 acres) and was supported by nine masts each 80m (262ft) high from which hung a network of cables.

The Millennium Dome at Greenwich, covers around twenty acres and is considered to be the largest domed structure in the world. If it is treated as a tent, it will be small compared to a much bigger one covering 105 acres in Jeddah to shelter the pilgrims in Holy Muslim sites. It is said that the first great dome built was for the Pantheon in Rome around AD 123. Spanning 43m (142 ft) barely a seventh of the Millennium dome, it was built of concrete and still stands. The Roman builders knew that a dome would bulge at the bottom if not restrained and built the supporting walls 7m (23ft) thick. Later builders and architects such as Christopher Wren at St Paul's Cathedral ringed their domes with iron hoops or chains. The Greenwich Millennium Dome is roofed with very strong translucent fabric and it is unlikely to stand for as long as the Pantheon. Some predictions give the fabric a life of 25 years but others believe the dome will, with maintenance and replacement, last for over one century.

The giant dome forms the centrepiece for the Millennium Exhibition at Greenwich in south-east London. The original design was submitted when a formal application to Greenwich Council on the 53 hectares (130 acres) former gas works. The dome was designed by Sir Richard Rogers and his architects, Imagination, the exhibition designer, and engineers, Buro Happold of Bath and London. The intention was to house twelve multi-level pavilions each dedicated to a different aspect of time. Surrounding the structure which has spectacular views over London and the Thames, are parks, gardens and entertainment amenities, with transport facilities for boats and the new Jubilee Line underground extension.

The 90m steel masts were being assembled in September 1997 across the Thames from Canary Wharf and they gave the first clues to the vast scale of the Millennium Experience project. A complex web of stay cables and tie-downs was developed to stabilise the masts. These take the weight of the fabric cladding in a supporting network of radial string cables and circumferential cable trusses through a pyramid of hanger cables. In all 26 cables have to be attached to each masthead and 18 at each base. At the centre of the roof a 30m (100ft) diameter 3-dimensional cable structure eliminates potential sagging. The bright yellow paint for the masts replaced dull grey when the design life of the structure was extended late in 1996.

For durability reasons it was decided by the masts contractors to use welding rather than bolting for final assembly on site. The contractors had to produce a minimum of one mast a week to satisfy delivery time and programme of construction. In October 1997 the primary cable network was knitted together at ground level. The mast raising began the same month when a 600-tonne crane joined another 200-tonne crane on site for a series of twelve dual lifts in twelve days. Temporary cables had to support the masts while the cable network was gradually connected and lifted into the air. By February 1998 the structure was more or less complete with most of the cables in place.

The structure is suspended from a dozen 90m (300 ft) high steel masts, held in place by more than 70km (44 miles) of high strength steel cable. The roof fabric is self-cleaning specially coated glass fibre which should last at least twenty-five years and can be renewed in sections if necessary. The two-layer fabric provides insulation to prevent condensation. Construction started in June 1997 with the driving of the first piles into the site. The twelve masts were erected in October and work started on knitting together the steel cable net to provide the tensioning and support for the Dome's masts and roof fabric. Work on installing the roof panels started around May 1998.

During the tendering for the Millennium Dome roof Greenpeace protested about the materials used. The Government Minister responsible decided that the roof should be made of Teflon-coated glass fibre instead of PVC as this is more environmentally friendly and longer-lasting. An American firm won the new £14million contract.

The Dome Roof

The Millennium Dome roof defied all critics when completed ahead of schedule. The Prime Minister, Mr Tony Blair, was guest of honour at the 'topping out' ceremony of the Dome on Monday 22nd June 1998. Television coverage of the event was seen by millions of viewers in the United Kingdom, Europe, United States, Asia and indeed all over the world. The first panels of the fabric began to go up on 23rd March, slightly earlier than originally forecast and the roof rapidly took shape. The roof is the largest of its kind in the world and has been submitted for entry into the Guinness Book of Records. The team of abseilers working at a height of up to 46m (150ft) above the ground could be seen from many parts of Docklands and Greenwich.

Life of the Dome

The Dome organisers New Millennium Experience Company (NMEC) have stated that the Dome structure could still be around in the 22nd century and after. It was assumed, conservatively, that the roof structure definitely would last twenty-five years, simply on the basis that this is how long the oldest similar structure has been around. However, the roof panels are designed to be renewable so there is no reason why, with proper maintenance, it should not be around for hundred years or more. The 8 hectare (20 acre) roof is one of the largest in the world. With one hundred thousand sq.m (1million sq.ft) of Polyester the roof could support the weight of a Jumbo Jet! It would take 3.8 billion pints of beer to fill the Dome. It could also contain two Wembley Stadiums, ten St. Paul's Cathedrals, the Great Pyramid of Gaza or the Eiffel Tower on its side. During construction the Dome's roof needed heat sealing and awaited delivery from the manufacturers of eight panels to be installed around the ventilation shaft to the Blackwall Tunnel in one corner of the Dome. Work on the central arena, lifts, escalators and foundations started in August 1998, while installation of the attractions within the Dome was scheduled from October.

Suggested future uses for the Dome after the exhibition included a film studio, a conference centre or sports arena to boost London's bid for the 2012 Olympic Games. The Prime Minister said the Millennium Dome is too good to knock down and its builders claim it could be around for centuries providing a lasting legacy beyond the year 2000. Even at this very early stage, a number of proposals and enquiries have been received. However, the Dome's long term future can only be determined by the site owners, English Partnership, after the year 2000.

This picture shows a construction supervisor at 100m (330ft) up a ladder attached to one of the slanting masts of the dome. Both the supervisor and the photographer, Andrew Baker, were safely secured to the ladder and glad to come down to earth!

A glimpse of the inside of the dome after completion of the roof in June 1998. The dome is 50m high (164ft), the same height as Nelson's Column in Trafalgar Square and big enough, 8ha (20 Acres) to contain 13 Albert Halls or 2 Wembley Stadiums.

Aerial view of London Docklands and computer image of the 320m (1050ft) diameter Millennium Dome at Greenwich for the celebrations in Year 2000 and beyond.

Assembling on the ground the central cable ring for the dome roof. When completed more than 70 km of cable supported the self-cleaning Teflon-coated cladding.

The last of the huge fabric roof panels are unfurled while other teams fix panels to cables. Trucks trundle across the floor beneath them, and the first service pipes are positioned, circa May 1998.

Millennium Dome in statistics: 1 km in circumference, 25 year minimum lifespan for roof fabric, 50m high at its centre, 100m high masts, 8000 piles, 30,000 visitors every day and 12 million visitors during year 2000.

Construction of Millennium Exhibition Dome in London

International Millennium Celebrations

World-wide Celebrations

Every day another group, another country jumps aboard the Millennium bandwagon. **Iceland** lights bonfires. **England** has a nation-wide pealing of bells and maybe a bash at **Greenwich,** the home of Time itself. British Airways has a giant Ferris wheel that lights up the **London** skyline. **Sri Lanka** holds a gigantic beach bash in Colombo on the shores of the Indian Ocean. **Brazil** has a party in Rio de Janeiro, on the shores of the Atlantic Ocean.

London with its stranglehold on Time may claim to the world's major millennium celebration - but what happens at Greenwich won't be the first, nor the last, celebration of the 21st Century and the Third Millennium. There are mega events planned for one million plus people in every continent. The UK has led the way in terms of celebrations for the year 2000 with the USA, Germany, France, Italy, Hong Kong, New Zealand, Canada, the Holy Land and South Africa close behind.

France is celebrating with the theme of 'France, Europe, the World; a new inspiration'. Its events include the construction of a 12km walkway along the river Seine, the placing of 2000 giant metal fish extending three metres above the surface of the river and the reopening of the Pompidou Centre.

Holy Land 2000

The Holy Land is a major focus for religious celebrations marking the millennium. Events include: a divine service in Jerusalem on Christmas Day 1999 and 2000; scientific theological conferences; religious music concerts and a wide-ranging programme of events entitled 'Bethlehem 2000', the centrepiece of which is an infrastructure programme to help Bethlehem cope with the likely influx of visitors. A total of four million visitors are expected to stream into the Holy Land during the year 2000, to commemorate the anniversary of the birth of Christ. Special events are being hosted in Bethlehem, Nazareth and Jerusalem from 1999 to 2001.

Vatican Holy Year 2000

From Christmas 1999 to 2000, the Vatican expects more than 13 million tourists to visit the basilicas of Rome in order to celebrate the Great Jubilee of the Incarnation of Christ as declared by Pope John Paul II and the advent of the third millennium.

Millennium March 2000

In honour of the 2000th anniversary of the birth of Christ, more than 30 million people are expected to participate in a global "March for Jesus" celebration to be held around the world on June 10, 2000.

Earth Day 2000

All around the world on April 22, 2000, more than 300 million people in 150 nations are expected to participate in the largest Earth Day event in the year 2000. It is also the 30th anniversary of Earth Day.

Hanover Expo 2000

Germany has major events such as Expo 2000 in Hanover 1st June-31st October, 2000; a programme of celebrations entitled 'Berlin 2000' and privately financed millennium festivities taking place in several cities.

USA Cultural Events

The USA has themes of preserving US culture and history, encouraging creativity and exploration and extending the bounds of international understanding. Events include: Millennium on the Mall – a folklore festival featuring children from around the world; Millennium Communities – a programme to encourage local projects to mark the new century and Millennium Minutes – a series of 60-second TV shorts by the National Endowment for the Humanities on the USA's cultural heritage.

International Centre for Life

The American razzmatazz that swept Universal Studios and the Smithsonian Institute to visitor success has been earmarked for the £54million Millennium spectacular attraction built in Newcastle upon Tyne, England.

The city's landmark project, the **International Centre for Life** celebrates the greatest scientific discovery of all time - the genetic recipe which makes every man, women and child in the world unique. And to ensure it pulls in the visitors from the opening in 2000, **Tyne and Wear Development Corporation (TWDC)** have hired **Rouse Wyatt Associates** from Cincinnati as master planners. The traditional lines of demarcation between museums and exhibits, and theme parks and entertainment are fading. Rouse Wyatt's approach of using strong storylines and immersive environments to attract visitors, then engage them in the joy of learning, is something which echoes our own thoughts.

The 4 hectare (10 acre) site also houses two other elements of "Life" in the form of a **Genetics Institute** and a **Bio Science Centre,** creating a "world first" for Newcastle that is expected to attract 300,000 visitors a year.

City Celebrations

New York Times Square transforms into a kind of cultural cross-roads on 31 December, 1999, with giant television screens conveying millennium celebrations from each of the worlds 24 time zones.

More than one million people are expected to meet for New York's Millennium celebrations on the New Millennium Eve on 31st December 1999. It was planned with a practise run in December 1998 when a similar number of people turned out.

There's no dispute that the new millennium actually begins in the year 2001, acknowledges the co-chairman of the **Washington** based Millennium Society. "The dynamics of throwing a party for planet Earth a year early include the magic of seeing the number 1999 turning into 2000 and launching a year of events and celebrations."

Boston is enjoying a brand new £1.4million leisure complex for the Millennium with a giant 4,700-seater Sony multiplex cinema and Reebok sports club at its centre. The centre, named Millennium Place, brings life and energy to the city centre. A group of Yale graduates founded the Millennium Society in 1979 to plan "the largest charity fund raiser in the history of the world."

Beneath the murky waters of the River Liffey in **Dublin,** a giant digital clock is ticking down the seconds to the third millennium The clock erupts in fireworks at the stroke of midnight on the last day of 1999.

Even though technically the new century does not begin for another year, the lure and symbolism of the year 2000 is much too seductive for the world to ignore. Plans include a round the globe succession of black tie parties and concerts at historic sites, including **Mount Fuji** in **Japan,** and The Great Wall of **China** and the pyramids in **Egypt.**

Beginning of 21st Century

Greenwich, **England,** which is used as a basis of standard time throughout the world, claims the new century officially begins on its stroke of midnight. That argument has few takers in time zones nine hours ahead, where the competition is frenzied to be "first". The Greenwich countdown clock has been joined by two other clocks, ticking their way to the Millennium in Paris.

A **New Zealand** based Millennium Adventure Company has already secured the rights to the worlds "first light" on Pitt Island. A motel on the island says it has had plenty of inquiries, but has not taken any bookings yet.

On the East coast of **New Zealand's** North Island, meanwhile, the city of **Gisbourne** is organising a Mardi Gras festival and billing itself as the "First City in the World to see the Sun." Other islands, including the Kingdom of **Tonga** and the Republic of **Kiribati** in the South Pacific, are also pitching themselves as the first places on the planet to usher in the next 1,000 years.

Fiji has focused around the fact that it is one of the first countries to see the sun rise on 1 January 2000. The theme is 'Tomorrow begins in Fiji' and includes a two week International Festival of Music and Arts; an interactive technology expo and a millennium wall built on the 180th meridian which is made up of hollow bricks with millennium messages inside which people can buy.

Just west of the date line, Western and American **Samoa** are banking on their own share of millennium madness. As the last places on the earth to dust off 1999, they're hoping tourists will come to the islands not to celebrate the new century, but to bid farewell to the old.

First Light for Third Millennium

New Zealand has a time vault which is opened at the start of the fourth millennium, the installation of the world's biggest tuned bell in the national Carillon in Wellington and an international children's festival.

The Year 2000 Board of **New Zealand** is to prepare celebrations for the new Millennium dawns. New Zealand has claimed that because of its position on the date line and its high altitude, it is the first country to usher in the new Millennium. They claim they are in an unrivalled position to attract tourism and world notice as the first nation in the world to welcome in the new Millennium. Arrangements have been made to tie in with the Sydney Olympics and the America Cup in Auckland.

Years of scientific research has shown that the first inhabited place to see the first Millennium light is Mount Habika on Pitt Island, just east of the international date line, about 745 miles Southeast of **Christchurch.** The sun hits the island which is 370 miles east of New Zealand at 3.59am, which is equivalent to 14.59 Greenwich Mean Time.

Religious Celebrations

Christian Anniversary

The Millennium is a Christian anniversary around the world, by the convention two thousand years since Jesus Christ's Birth. The Pope declared the year 2000 a Holy Year or Great Jubilee for the Roman Catholic Church. Thirty five million pilgrims would be expected in Rome during the year. The years around 2000 are seen as genuine renaissance of creativity and imagination, and this renaissance has already begun.

In this country, with its long Christian history, the churches have planned to mark the first weekend of the year 2000 with a number of special services. There would be a nation-wide peal of bells at mid day on New Years Day. They have also decided that the Pentecost weekend in June that Year would be a main focus of their celebrations. The Millennium period is used as a time for reflection and renewal both within the churches themselves and in their commitment to the wider community.

Close consultations took place with all the mainstream, Christian denominations over how the centrality of the Christian message may best be reflected in the celebrations. A meeting was held at Lambeth Palace with representatives of the churches and members of a number of other faiths' communities. The volume of work the churches put into planning for the millennium has been impressive, and also by the depths of inter denominational co-operation.

The celebrations surrounding the Millennium are being welcomed by members of many different faithed communities with whom the Millennium Commission were in regular discussions. To these communities, the Millennium offered a chance to rejoice in and reflect on so much of what we all share in common. This coming together is based on a sharing of common values and a welcome for diversity, not on an artificial pooling and dilution of different religious beliefs.

Whether you approach the Millennium from a spiritual or secular perspective, the idea of renaissance - of regeneration and renewal - can be shared by all. Discussions took place between the Archbishop's Millennium Advisory Group and other faith communities, which resulted in a paper on the Millennium and other faiths. This document was published together with a list of aspirational values for the Millennium Exhibition, prepared by the Churches together with the England Millennium Planning Group. This identified moral values with which many of us will be able to associate.

Regeneration and renewal are not exclusively Judaeo-Christian messages. As Indarjit Singh pointed out in a 'Thought for the Day', the great Sikh Gurus constantly emphasised the essential unity of religion and the commonality of core values essential for sane, balanced and responsible living in any age. These values are universal, transcending barriers of religion and political allegiance. They have been underlined by HRH the Prince of Wales, whose contributions to the debate about the Millennium are warmly acknowledged. They apply not only to individuals but also to our physical and social fabric. The Millennium may cause people to reflect about their lives and make resolutions. There will be a greater degree of reflection and taking stock - both personal and communal. Let us provide for this too.

An idea was put forward by the Churches, for everyone to have a simple, small candle, which they would light - wherever they are, in their homes or streets, or at celebrations - during a brief period of silence just before midnight on New Year's Eve 1999. This would be a 'Minute for the Millennium'; a chance for us all to pause and take a breath before we embark on the great adventure of a new century and a new Millennium. A shared moment of stillness, and of prayer for those of us who wish to pray, is not too much to ask for or to encourage.

There are other ways in which people can be brought together; it is rooted in Scripture, specifically the Books of Daniel and Revelation. But the concept of the Millennium has an alternative scriptural provenance, that contained in Leviticus. This views the Millennium as a greatly enhanced version of the tradition of Jubilee, the Hebraic year of redemption and forgiveness which recurred every seven years. Jubilee entailed the forgiveness of debts, the righting of wrongs and the renewal of society.

The Roman Catholic Church has termed 2000 a year of Jubilee. The terminology pointed to a fresh start. We are a fortunate generation in that we have before us the unique opportunity of the transition to the Third Millennium to put things right. It is a time for optimism, a time for regeneration. It presents everybody with the chance to build bridges, and the chance to renew our cultural traditions of respect, tolerance and service.

Christmas Celebrations

In Stuart times, people loved Christmas with its religious services, dancing, theatre performances, high jinx, eating and general making merry. Homes from the grandest palaces to the humblest hovels, were bedecked in evergreens in a bid to ward off evil spirits, brighten up surroundings and, in the case of mistletoe, to provide a good excuse for a kiss and cuddle. Another long-term favourite was to 'Lord of Misrule'. A lowly member of the household would be elected as King for a day and would preside over a host of indoor tournaments and games. Christmas was also a feast day and food played a prominent role. Poultry became highly fashionable and expensive. The ritual of exchanging Christmas gifts did not develop until the Victorian era. The King and Queen would give charitable gifts to their poorer subjects.

Just as today, men, women and children, would dress in their best clothes for Christmas Day. Men often dressed in fashionable breeches that were so full they looked strangely like skirts. A recent demonstration of life during the reign of Charles II was shown by the Past Pleasures company at the National Maritime Museum, Greenwich.

Westminster Abbey

Adjacent to the Houses of Parliament is Westminster Abbey which houses the Coronation Chair. The Gothic style Abbey has been the place of crowning English Kings and Queens since Edward the Confessor, who built the original church in 1065. There are tombs and memorials of past Kings and Queens and famous British subjects over the centuries. Among the Royalty and Statesmen buried here are Queen Mary, Queen Anne, Pitt the Elder and the Younger, Charles Fox, who fought to abolish slavery, Lord Palmerston and Gladstone. Famous scientists and engineers who have memorials include Sir Isaac Newton, Lord Rutherford, Lord Kelvedon, Michael Farraday, Robert Stevenson and Thomas Telford.

St Paul's Cathedral

The original Cathedral was founded in 604 and rebuilt in stone by the Saxons later in the same century. Further rebuilding took place in the 10th and 11th centuries. After the Great Fire of London in 1666, Sir Christopher Wren undertook its complete rebuilding between 1675 to 1710. The Gothic Cathedral was designed round a central lantern about 111m (364ft) high with elegant twin towers and an extensive vault. It contains many externally sculptured details and wrought-iron screens. It is based on St Peter's in Rome but on a significantly smaller scale.

Southwark Cathedral

Situated at the south-western end of London Bridge, Southwark Cathedral is one of several great Cathedrals in London. The first church was built during the 7th century on the site of a Roman Temple. There are Roman tiles in the pavement at the entrance to the south choir isles of the present building. Early in the 12th century a Norman Priory Church was erected and called St Mary Overie, which meant over the river. The birth in 1607 of John Harvard, who founded Harvard University in America and his baptisms at St Saviour's, is commemorated by Harvard Chapel.

Westminster Cathedral

In Francis Street, Victoria, is the most important Roman Catholic Church in England. Designed by Bentley and built 1895 to 1903, it is a wonderful Byzantine style Cathedral depicting early Christian architecture. The interior of the church is covered with mosaic set tiles and is ornamented with more than a thousand different kinds of marble from quarries all over the world.

City of London Churches

The City has fine historic churches many of which were designed by Sir Christopher Wren. Currently there are over forty churches which can be visited within the square mile. Several of these churches are dedicated to a particular cause or aspect of church work but sadly some are scheduled for closure. All Hallows by the Tower was built by George Dance the Younger in 1767 on the site of a much older church from Roman and Saxon times.

St Alphage Church in Greenwich

The 17th century masterpiece by Hawkesmoor and is one of six built by him primarily in East London. Following their raid in Canterbury in 1012 the Danes bought St Alphage to Greenwich and murdered him on the site where the church stands. The church was restored following the bombing of World War II.

St Marys Church in Rotherhithe

Built by local shipbuilders in 1715, close to the underground station, on the remains of an older church of Saxon origin, it features tree-trunk pillars and a barrel shaped roof structure similar to the hull of an upturned ship. It overlooks the River Thames.

Visit to the Holy Land

In December each year many people visit the historic treasures of the Holy Land and Jerusalem – home to the Wailing Wall, Dome of the Rock, Temple Mount and Way of the Cross. For a breath-taking view of Old Jerusalem visitors climb to the top of the Mount of Olives, dotted with tombs, crypts, monasteries and the Garden of Gethsemane.

Westminster Abbey, nearly 1000 years old.

Aerial view of the Houses of Parliament with Big Ben, and Westmi

Lord Mayor's Show along the Thames, showing St Paul's, circa 1747.

**St Martin's Church
at Trafalgar Square.**

**Choir, Sanctuary a
Southwark Cathed**

St Mary's Church, Rotherhithe, built 1715.

Westminster Cathedral, Victoria Street.

The Choir in St Pa

bey.

Aerial view showing Westminster Cathedral.

A view of St Paul's Cathedral completed 1710.

p Fox's Screen,

The Shakespeare Memorial,
Southwark Cathedral.

All Hallows Church near Tower of London contains Roman and Medieval remains.

hedral.

Chapel at Great Ormond Street Hospital.

LONDON PLACES OF WORSHIP

London Central Mosque, incorporating a steel dome with copper finish, in Regent's Park, North London.

The Hindu Mandir at Neasden, North London, built in limestone and marble.

A 19th century painting of the Spanish Bevis Marks Synagogue in the City of London.

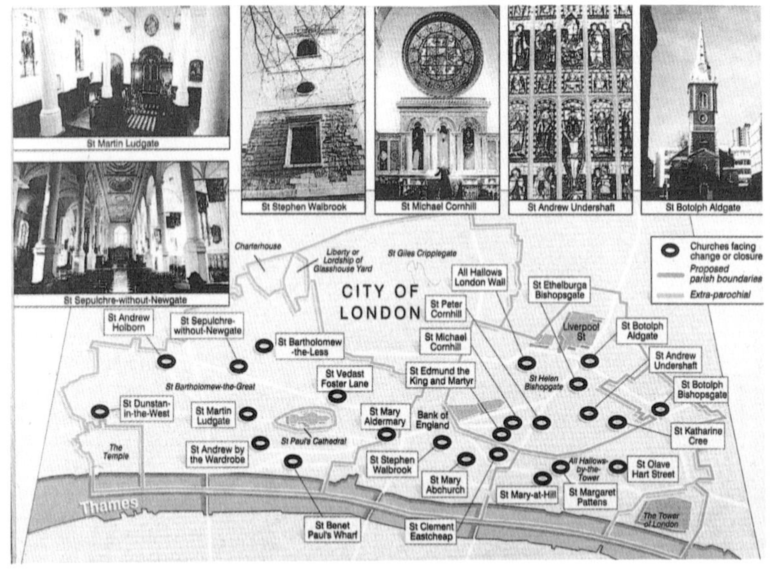

The City of London Churches, many of which were built in the style of Sir Christopher Wren in the 17th and 18th centuries.

Japanese Peace Pagoda in Battersea Park, South London, erected in 1985.

OTHER PLACES OF WORSHIP IN LONDON

London in the First Millennium

Prehistoric London

At the east end of Chiswick Mall, West London, there are below the top soil of 0.3m, 12m of gravel and sand, and below that 42m of Blue London Clay. At the western end, a short distance north of Kew Bridge, on the Thames, a deep cutting for the London and South Western (Windsor) Railway in 1838 yielded a remarkable discovery of mammal remains in a river drift deposit of sand and gravel, about 2m thick, lying just above the London clay. This included, among boulders of quartz, granite and rock with ammonities, the bones of an elephant, rhinoceros, hippopotamus, short-horned ox, red deer, reindeer and even the great cave-tiger, the lion, of which a well preserved foreleg-bone was identified. The bones are recorded as not waterworn, implying that these animals wandered here or prowled close by the river as it spread extensively along the valley.

Further west, in 1813, were found the teeth of elephant and hippopotamus and an elephant tusk nearing 3m and below them, in the blue clay, fruit and marine fossils, including crabs and oysters. The numerous and scattered relics from the Palaeolithic, Neolithic, bronze and iron ages, which have been in the Thames river bed, suggest human settlements up to the time of the Roman invasion. Similar finds were discovered in 1997 further upstream in the Thames Valley near Oxford; the elephant tusk and lion bones were dated about 20,000 BC. Sharp stone tools and flint axes were also found indicating human presence in the area of the same period. The magnificent Iron Age shield found in the River Thames at Battersea (see page 25), is made of bronze and inlaid with red glass and may have been used as an offering to the pagan gods.

When Julius Caesar, the great Roman General, landed on the coast of Kent towards the end of August, 55BC, his progress was impeded by a band of brave native warriors who long before had acquired some skills in agriculture, mining and commerce. They had also mastered the art of navigation of rivers for transporting and carrying goods. The people had a chief for a king whose rule was obeyed and whose dominion extended over the lands and pastures where his subjects produced their crops and tended their animals. The ancient Britons worshipped the serpent, sun, moon, fire and water.

Roman London

The Roman occupation of Britain which lasted from 43 to the year 410AD was in general beneficial to the country. The Emperor Claudius visited Britain for a short period at the beginning. There were many generals who ruled the country including Julius Agricola from 78AD about whom his son-in-law Taeitus, the great Roman writer, wrote many documents. Agricola invaded the northern part of the United Kingdom, known as Caledonia, and succeeded in building a line of forts and walls.

The Britons copied many Roman habits, the men of London and York adopted the toga as a fashionable garment, temples, halls and theatres were erected and comfortable dwellings of wood, brick or stone superseded the old mud huts formerly occupied by the people. Numerous baths were constructed, an excellent example of which can still be seen at Bath or Aqua Sulius, as it was then known. In this way the Roman law, religion, learning and arts were gradually adopted.

To enable their troops to move rapidly from one part of the island to another, the Romans constructed excellent roads composed of successive layers of stone, gravel, lime and concrete. The traces of many of these form the lines of major roads and motorways of today. The Emperor Hadrian, who visited Britain in 121AD, built a famous wall from the mouth of the Tyne to the Solway Firth. Many portions of this wall still exist today and can be clearly seen near Halstead, Northumberland and Chalford where the line is intersected by the Tyne and where there are the remains of a military station called Chesters.

The greatest benefit conferred on the country by the Romans was the establishment of settled law and order, under which the various existing communities developed. Men no longer settled their quarrels by violence but in the Law Courts where the Romans proved themselves to be perfect masters in the art of doing justice. It is a remarkable fact that the study of Roman Law still today forms an important branch of reading for all who desire a career in the legal profession.

Anglo-Saxon Times

The departure of the Romans left the Britons as easy prey to the Saxon warriors who invaded the island during the 5th century. The land was rich in corn pasture and animals and the Saxons settled and took over the countryside rather than settle in large towns. Doubtless many of the men among the Britons were killed but some remained to till the land for their new masters.

Following the withdrawal of the Romans, c410, the country was divided into a number of kingdoms. Vortigern reigned a large kingdom from about 425AD and he used Saxon mercenaries, led by Hengist and Horsa to defeat the Picts. But c441, the Saxons revolted against him and established a kingdom in Kent which extended to London. Subsequently, many more Anglo-Saxon settlers arrived from Germany. By 600 the Anglo-Saxons had permanently mastered most of England. In his Ecclesiastical History of the English Nation, Venerable Bede provides the main source of knowledge of the preceding centuries.

In the early Saxon days the people worshipped many Gods, one of them being the Viking War God, Odin, from whom we possibly derive our word Wednesday, or Wedin'sday. He was depicted as one-eyed, since he had bartered the other eye for wisdom. Thor the God of Thunder, or the Air, gave his name probably to our Thursday; whilst the names of Teu, the God of Darkness, Freea, the Goddess of Beauty and Soetere, the God of Hate still survive in our Tuesday, Friday and Saturday. Sunday and Monday were respectively the days of sun and the moon.

No contemporary documents or maps exist to support the archaeological findings of the living places of Saxons mostly because their buildings were made of timber. Nevertheless, evidence that has been discovered shows that this was a prosperous period in the post-Roman years. It would seem likely that many of these Saxon areas are those that remain preserved in parish boundaries even today.

In 601AD St. Augustine arrived in Kent from Rome on his mission to convert the Saxons to Christianity. The process was largely complete by 700 and Christianity bought to an end though not necessarily immediately after introduction, the Pagan tradition of burials at sites in many parts of London and the country. Late Saxon England also saw other changes, which laid the basis for much of our development. These include the emergence of widespread trade within a money economy and the growth with it of thriving towns. This was particularly the case in coastal areas and in London. Many present-day towns such as Guildford and Southwark emerged as towns before the Norman Conquest.

The Sutton Hoo Treasure

In 1939 archaeologists found at Sutton Hoo, on the River Deben in Suffolk, a buried ship loaded with treasure. It is believed to be a cenotaph to an East Anglian King. The hoard of gold coins indicated a mid-seventh century origin. The treasures included some magnificent examples of Anglo-Saxon workmanship, which can be seen in the British Museum, London such as gold clasps inset with garnets and glass and a gold purse decorated with inlaid garnet and mosaic glass.

The Vikings

The period prior to the arrival of the Normans also experienced Viking raids as well as Saxons long term warfare with the English kingdoms. As a result unlawfulness increased along with rising prosperity and the late Saxons Kings sought control by means of increasingly rigid laws. There is some evidence that burial grounds show victims of a massacre of the Danes fleeing from a defeat in London in 851.

The ancient Norwegian word Viking means pirate. The success of the Viking expeditions to London and England depended mainly on their skills as ship builders and navigators. The myth of the Viking wearing a helmet with horns is well known but they also wore pointed hats or iron helmets. Many Viking expeditions sailed from Sweden for the purpose of plunder and trade with England during the 8th and 9th century. Many of these Vikings settled in London, the Midlands and the north of England.

As we enter the third millennium the people of the United Kingdom may be excused for thinking that they are sitting in the middle of a fast lane to Europe. The ancient English and Saxons probably felt similar insecurity but to them the consequences of change were probably far more brutal!

Recent Excavations

One of London's richest archaeological sites was discovered during excavations (1995/96) for the Jubilee Line Extension in Borough High Street, Southwark. Archaeologists have unearthed thousands of Roman artefacts including imported pottery, coins and remnants of everyday life. The site, which is the new ticket office for London Bridge Station, was once one of the two sandy islands in the Thames just off the bank of Southwark. The Romans filled in the channel between the two islands around 200AD and built a road running close to today's Borough High Street and leading to the wooden London Bridge constructed earlier by the Romans.

At Westminster Station, excavations found evidence of occupation during prehistoric and medieval times. Buildings found included an inn named the Saracens Head, a gatehouse and a merchant's house in the middle of Parliament Square. At Stratford market depot, a large medieval Cistercian Abbey dating back to the 12th century, was found complete with burials of monks over 400 years old. The site also revealed almost unknown Iron Age and Roman remains.

Activities near the Roman warehouses and timber bridge, almost on the site of the existing London Bridge.

The Thames and Londinium from the north-west, showing the Roman city wall, circa AD200.

Typical rural village life early in the first millennium
The temple on the left was also the village hall.

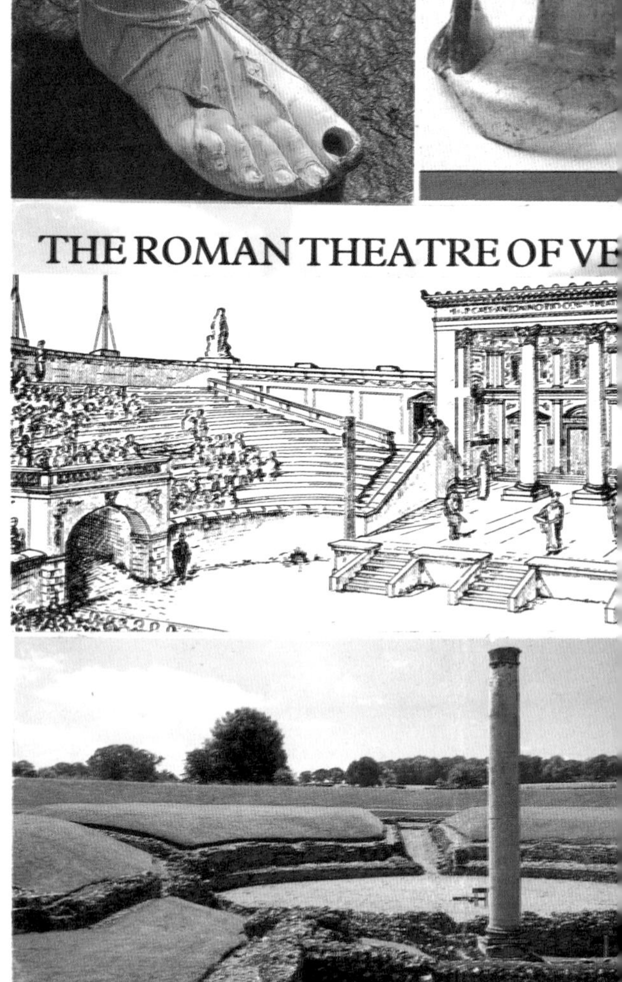

Top: **Roman oil lamp, jug and pottery, found at Lond**
Bottom**: Impression and remains of the Roman Theat**

Glimpse Of Saxon Past

First Millennium

pression of landing of the first Anglo-Saxons
nt under their leaders Hengist and Horsa.

Artist's impression of a typical Saxon village in Surrey, South London, showing timber buildings and funeral procession outside the village.

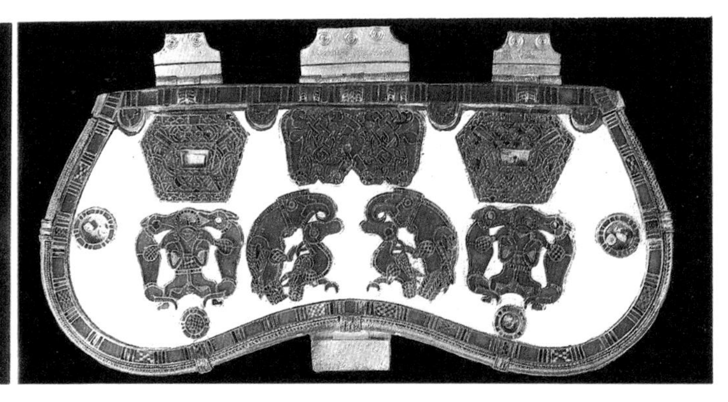

A gold purse decorated with inlaid garnet and mosaic glass, found in East Anglia.

Anglo-Saxon gold clasps inset with garnets and glass, found in East Anglia.

AMIUM

First Millennium

e.
lbans circa 100AD.

Impression of a Viking pirate ship used for raids on England, 800-900AD. Their success depended on these excellent ships.

A pre-historic shield found in the Thames at Battersea.

London in the Second Millennium

The Norman Period

Throughout the Saxon period, the power of the 'Lord' over all dependants had been gradually increasing. With the arrival of the Normans in 1066, this power further increased and the Manorial system was established. As soon as William the Conqueror established himself, he announced that the people would be allowed to use the land as 'held' from the King. Nobles therefore held the land directly from the King and each holder became the 'Lord of the Manor'. In 1083 the King decided that taxation would be necessary in order for the government to carry on working efficiently.

Commissioners were appointed to go around the country and collect information for the Great Survey and a mass of information was gathered in about nine months. This was carefully collected and written down by trained Norman clerks. The statements referring to each county were then sent to Winchester, then the Capital, and arranged in order of importance, county by county, to make up the wonderful old book known as 'Domesday Book'.

This remarkable book was referred to as the 'Grand Roll' or the 'King's Roll'. It is thought by some that the present title of the survey was derived from Domus Dei, the name of the vault in which it was first deposited in the Cathedral at Winchester, and affectionately sometimes referred to as the Roll of Winchester. The survey is in two volumes of which the second is the smaller. The volumes give a wonderful description of the social life of the Manors in the 11th century and are now kept in London. For a hundred years after their arrival, the Normans developed commercial trade in England and rebuilt old churches or erected new ones in many parts of the country, and many of our present Cathedrals, Abbeys and Churches bear some important Norman work. Norman castles and other strongholds are still to be found in various parts of the country, the Tower of London and Rochester castle being excellent examples.

The Crusades

The occupation of countries in the Middle East by the Seljuk Turks at the beginning of the second millennium provoked a series of wars, the Crusades, for the recovery of the Holy Cross and Jerusalem. The crusades to the Holy Land played an important role in the life of western Europe from the late 11th century. The First Crusade was ordered by Pope Urban II in 1096 and it led to the recapture of Jerusalem and the massacre of many prisoners. The Second Crusade under pressure from the King of France attempted unsuccessfully to recover the City of Edessa. In 1187 Saladin captured Jerusalem.

The Medieval Period

The Medieval period covers the years between the accession of King John in 1190 and the death of Richard III in 1485; a long period during which many changes took place and by the end the country had become a nation of great power. The first mayor for the City of London, Henry Fitz Aylyain was appointed during the reign of Henry III and remained in office for twenty years until his death. Much of London's wealth could be seen in the bustling trade of its port, which by 1400 was one of the largest in western Europe. The port's commercial prosperity came about largely from the activities of trading guilds and associations as Merchant Adventurers.

Between 1176 and 1209 the old stone London Bridge was built by Peter De Colechurch who was a Chaplain of the church of St Mary Colechurch. Later houses and shops were built on the bridge along with a chapel dedicated to St Thomas Becket, where people crossing could pray for safe passage and hopefully leave a donation towards maintenance. A fortified gatehouse secured the southern end in Southwark where heads of traitors executed on Tower Hill were exhibited. There was also a drawbridge and for ships to reach the upriver wharves, this had to be raised at a toll of sixpence.

By the Magna Carta of 1215 King John granted the City a Charter which confirmed the right to chose a mayor annually. The Lord Mayor yielded considerable powers second only to the monarch of the time. The title 'Lord Mayor' came into being in 1414. One of the most famous mayor's of London is Dick Whittington, born 1360, who held three full terms.

Generally the traders in the City were divided into a number of guilds and craft guilds and great rivalry existed between them. In terms of age the Merchant Tailors established in 1326, is the oldest, followed by the Goldsmiths and Skinners, 1327. By the 14th and 15th centuries, the Livery Companies were extremely powerful and participated in the government of the City of London and gained rights to elect the Lord Mayor and Sheriffs of London.

In 1348 Londoners experienced the miseries of the Black Death for seven months, killing some two hundred people a day. The Black Death also heralded the Peasants Revolt in 1381 headed by Watt Tyler, for with the death of so many those who were left realised the power they held to demand more wages and better conditions. It was Richard II who had a troubled reign and faced Tyler and his men.

Tudor and Elizabethan Ages

With the arrival of Henry VII, the medieval way of life came to an end and in its place the Tudor Dynasty started and lasted nearly 120 years from 1485 to 1603. At the beginning London was essentially Medieval, bounded by its old Roman wall within which most of its sixty thousand inhabitants lived, governed by the Lord Mayor and his Aldermen. The great houses of the nobility, merchants' gabled houses, mini markets and over twenty monasteries and some one hundred churches with St Paul's dominating the scene, all clustered together in the City. The growth of London resulted in an increase in population to 200,000 by the end of the 15th century, many of whom were no longer living within the walls. During this age many prisoners were executed at the Tower of London, amongst them Sir Thomas Moore, Bishop Fisher, Anne Boleyn and Lady Jane Grey.

Most Londoners were busily involved in trade and there was an influx of immigrants into the City, many fleeing from religious persecution on the Continent. Among them were Flemings, who were the first to put hops into their beer (greatly improving its quality), the French, Italians and the silk weavers who settled in Shoreditch and Spittlefields.

By the middle of the 16th century there were a number of Royal Palaces along the Thames. The Palace at Greenwich where Henry VIII was born, Hampton Court, built by Cardinal Wolsely but taken over by Henry VIII. Villages began to spring up around the City, among them Chelsea. Although in Elizabethan times Londoners were serious in their study of law and conducted international business, they also enjoyed having a good time. All sorts of sporting activities, including dancing and mock trials, took place on festive days, while marriages ended in great receptions and feasting. It was also the age of the birth of the English theatre in 1577 when James Burbage opened London's first playhouse called the Theatre in Shoreditch. It was so successful that many more theatres followed. Many of England's most famous theatrical names came from this era including William Shakespeare who's plays were presented at the old Globe Theatre which was built in 1599.

Soon after the return of Charles II to the throne the City was to be the scene of two disastrous events: the Great Plague and the Great Fire. The outbreak of plague in 1665 resulted in more than 70,000 deaths. On 1st September 1666 the Great Fire burnt the city practically to the ground. The medieval city had gone forever. King Charles II wanted a new London which would be graceful, neat and orderly and not so susceptible to fire and plague. He appointed a group of six Commissioners with Christopher Wren as their Head who worked day and night to plan a solution laying down building regulations for the first time.

Georgian London

Georgian London was an era of elegance. Canaletto painted scenes of the grace and beauty of London. At the same time William Hogarth depicted clearly the terrible squalor and poor conditions in which many common people lived. For most of the period of the first three George's England was at war with other countries. As these wars were carried out abroad, Londoners got on with their daily lives without any worries. During the Cabinet meetings held by George I and George II, the role of the Prime Minister began to emerge. This was an assistant post for the King who was the head of the state and chose his ministers. The first minister who's role was acknowledged by other Members of Parliament was Robert Walpole, while Pitt the Younger was the first to be actually called Prime Minister.

After the Great Fire of 1666, London needed new churches. At the beginning of the 18th century Nicholas Hawksmoor, a pupil of Sir Christopher Wren, was commissioned to build six churches. The East End saw three of them: Christchurch, Spittlefields, St Annes, Limehouse and St Georges in the East. In the City was St Mary Wall North, the remaining two were at St Georges in Bloomsbury and St Alphage at Greenwich. Living in the villages around London was made dangerous by attacks from highwaymen – Dick Turpin was just one who haunted the Hackney Marshes. This perhaps explains the fact the public hangings at Tyburn in Hyde Park, were one of the great spectator gatherings of the 18th century.

Second Millennium Pictures

The pictures on the next page span 700 years from the conquest of the Normans in 1066 to the 18th century, showing the colourful capital city and seat of England's Government. Medieval life in all its diversity is evoked through a number of objects that have survived in London's soil and been recovered by excavations. Saxon jewellery, Medieval leather bags, pilgrim badges are just a few of the hundreds of items that have been found and once belonged to Medieval Londoners. They are on display in the great London museums.

The Bayeux Tapestry was commissioned by Bishop Odo of Bayeux circa 1078, and illustrates events in the life of William the Conqueror.

Norman soldiers crossing the English Channel for the Battle of Hastings 1066. Note the horses being ferried across as cars are today.

King Henry VIII
at the age of 45.

The 15th century Medieval Guildhall
in the City of London.

Peter the Hermit preacher of the First Crusades, circa 1095.

The White Tower was built
by the Normans circa 1098.

Tudor ceremonial pavilion, circa 1511
and procession, circa 1520.

A vibrant Elizabethan wedding in Bermondsey,
near today's Butler's Wharf.

Queen Elizabeth I.

Riots in Broad Street, London, June 1780.

The Great Fire of London 1666. Note Tower of London on the right.

Oliver Cromwell ruled 1649-1660.

Second Millennium (1066 – 1780)

19th and 20th Century London

Victorian Period

During the reign of Queen Victoria from 1837 – 1901, London became the most powerful and wealthiest city in the world. It was not only the centre for banking, commerce and trade but also science, engineering and medicine. The industrial revolution had transformed the Midlands and the North but had not affected London a great deal. In 1837 the coming of the railways drastically changed that for they brought with them factories, manufacturing industries and an influx of workers. London's first railway ran less than four miles from Southwark to Greenwich. Euston Station became the first mainline terminus with trains to and from Birmingham. Greater London was soon criss-crossed with railway lines and many villages were linked with London. New communities were created but thousands of houses were demolished to make way for the lines. In January 1863 London saw the world's first underground, the Metropolitan Line, which ran from Paddington to Farringdon Street and within six months over 25,000 passengers were using it daily.

In 1840 Queen Victoria married Prince Albert, who was largely responsible for the Great Exhibition of 1851 in Hyde Park. Through the exhibition he illustrated the great might of Britain's artistic, scientific and commercial achievements. The exhibition was held in the Great Palace in Hyde Park spreading over 19 acres. The building was of delicate cast iron ribs and some 300,000 large panes of glass. From its opening by the Queen on 1st May, thousands came to see the exhibition which attracted 6million people in five months. It was a great success resulting in a profit which was used for the purchase of land to build the Victoria and Albert Museum, the Science Museum and Imperial College in South Kensington. The Crystal Palace was purchased by the Brighton Railway Company and re-erected at Sydenham where it remained until it burnt down in 1936.

Typhus and Smallpox killed thousands of people each year, especially in the poor areas of the East End. In November 1848 a cholera epidemic broke out which lasted eleven months. Worst hit were the areas near the River Thames. Sir Joseph Bazalgette, the Chief Engineer of the Metropolitan Board of Works, designed a new system for sewage disposal in the 1850s. Five main sewers, three north of the Thames and two to the south, carried the sewage over ten miles to Barking and Plumstead.

The River Thames continued to be used for commerce and transport. Steam boat services which started in 1816 ran four boats an hour to Greenwich. The trade along the Thames also continued to grow resulting in new docks being constructed: Victoria 1855, Millwall 1868, South West India 1870 and Royal Albert 1880.

In 1876 the Albert Memorial, which has recently been renovated, was erected in Kensington Gardens to his memory. The 15ft statue shows Albert holding a great exhibition catalogue, while in the tall spires are statues of those virtues so dear to the Victorian: faith, hope, charity, chastity and temperance.

The Victorians loved sports, thousands watched the annual Oxford and Cambridge boat race. The theatre was also popular and around twelve new ones were built in the West End. By the turn of the 20th century London had been completely transformed. Although traffic was still pulled by horses the motor car had made an appearance. The Prince of Wales bought the first Royal Daimler in 1900, motorised buses were running services and telephones were in operation. Improvements in living standards took place gradually, better street lighting and roads, libraries, museums and public baths were provided. Gas, water and finally electricity were laid on. Queen Victoria died on 22nd January 1901 but the Victorian way lasted another thirteen years before it was shattered forever by the First World War in 1914.

The Great Exhibition of 1851

The Industrial Revolution began in Britain during the last half of the 18th century. We were a trading nation that mastered the international business world and established a global market. By the time of the Great Exhibition our supremacy was total and visitors from all over the world were amazed at the range of products and equipment on display.

The exhibition provided as much excitement as Millennium Experience of 2000. The Dome which is the largest in the world has parallels with Crystal Palace which was a pre-fabricated structure. It clearly stated Britain's might in the world and featured much about the technological achievements of the middle of the 19th century.

London's Blitz

During World War Two, the main air attacks on London by Hitler's Luftwaffe started on Sunday 7th September 1940 in an attempt to destroy the docks and the bombing continued for 57 days. Past exhibitions of London during this period, have been both as a celebration and an attempt to recreate the terrible experience for new generations. Amongst the rubble, firemen worked despairingly and from the wreckage protruded dead bodies. At night and in the middle of an air raid, people lied huddled on bunks against the walls of underground stations.

People left the Underground and made for the sanctuary of Anderson shelters. Bombs exploded and sirens wailed in the skies above the flimsy metal roofs. Posters were everywhere: "Dig for Victory", "Join the Women's Land Army", etc.

When the war ended in 1945 Londoners turned their attention to restoring their capital to its former glory as one of the worlds most influential cities. A number of landmarks had been destroyed or damaged including the Tower of London, Westminster Hall, St Paul's Cathedral and the Guildhall.

The 1951 Festival of Britain

The 1951 Festival included the Dome of Discovery, the Skylon and many other attractions. The Dome of Discovery portrayed the achievements of British people in mapping and charting the globe, exploring the heavens and investigating the structure and nature of the universe. The Skylon was a dramatic silver pointer, which towered like a suspended exclamation mark 290ft above the ground.

The Festival was held to show new achievements in the Arts, Science, and Sociology. It was on the whole a great success. The exhibition was held on 27 acres of the derelict south bank of the Thames between Waterloo Bridge and County Hall. Many Londoners found enjoyment in the festival pleasure gardens laid out in Battersea Park. People queued for hours to buy commemorative five-shilling crowns as souvenirs. The South Bank has since developed into a centre for culture and enjoyment. The buildings including the Royal Festival Hall, the Queen Elizabeth Hall, the Puercell Room, the Hayward Gallery and the National Theatre were built as concrete buildings which are functional. Two years after the festival on 2nd June 1953, a million people lined the route from Westminster Abbey to Buckingham Palace for the coronation of Queen Elizabeth II, the present Monarch.

In 1967 a decision was taken to close the East India Dock. During the next fourteen years other docks were gradually closed and the traffic transferred to Tilbury. The great Royal docks were the last to close in 1981, many of the upriver wharves also closed at this time. Soon afterwards the empty wharves along the riverside were being refurbished into homes, offices and shops. Docklands other developments include the London City Airport and at the Royal Docks in Canary Wharf a new financial centre on the Isle of Dogs. During the second half of the 20th century London saw the growth of service industries and tourism. Travel by air meant that tourist could travel quickly to other parts of the world and visitors came to London attracted by its glorious history.

Heathrow Airport opened in 1946 and soon became a vital link for both business people and tourists. Later London was also served by Gatwick Airport in Sussex and Stansted Airport in Essex. Post 1950 also saw the development of new towns such as Stevenage and many council houses outside London. Today London has been transformed with the City and Docklands skyline of multi-storey office blocks. Old established trade markets have gone. Covent Garden moved to a new complex at Nine Elms and its old site turned into a fashionable shopping and café district.

Despite the changes, the 200 years of London's history still plays a vital part in its development into the new millennium. The annual Lord Mayor's Show for the City of London is still a popular event. The boat race along the Thames is held every year, as is the Doggetts Coat and Badge Race. The Tower of London, Tower Bridge, Madame Tussards, the great museums and art galleries attract millions of visitors each year.

London continues to grow and might swallow up everything around it until the south-east becomes one metropolis of offices, factories and housing developments. London has coped and adapted for 2000 years, there is no doubt it will cope for another 2000 years!

Regeneration of Docklands

Within the last three decades of the 20th century, the great London maritime trade that had brought wealth and life to the British Empire, ebbed away in the wake of the container shipping revolution. By the mid 1970s the historic landscape of docks and wharves became empty and desolate docklands.

Today these hundreds of acres near the heart of the capital represent one of the greatest challenges for urban renewal in the world. Already massive and dramatic changes have taken place, driven by private and public sector investments to bring new vitality to the banks of the River Thames. The most significant contributions to the renaissance and conservation are in the Isle of Dogs, St Katherines Docks, Wapping, Bermondsey and the South Bank, close to Tower Bridge. Canary Wharf, on the Isle of Dogs, is an impressive development to date.

Interior of the Great Exhibition of 1851, held in Hyde Park, London.

The Great Exhibition was housed in the 550m long Crystal Palace, erected in Hyde Park and relocated to South London. It burnt down in 1936.

Prince Albert Memorial in Kensington Gardens.

Queen Victoria reigned 1837-1901.

Westbourne Terrace, West London, circa 1830, a street of grand houses in Georgian London.

Bluegate Fields, Stepney. Overcrowded slums of Victorian London, circa 1870.

Early 1900 poster advertising house builders for development of suburban London

Bombing of London Docklands during the Blitz of 1940.

Royal Festival Hall built for the Festival of Britain 1951 and now part of the South Bank Art Centre.

Lloyds of London Headquarters, one of a number of modern architectures in the City of London.

London City Airport in the Royal Docks, opened 1987.

Dockland Light Railway, opened 1987.

The Shuttle Train at Waterloo Station for travel to Paris via the Channel Tunnel.

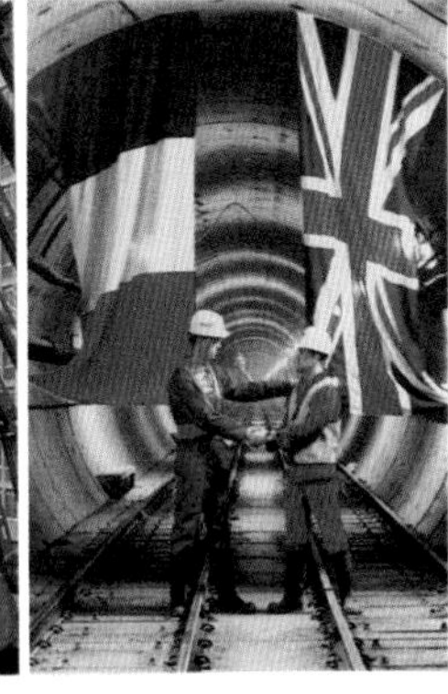

The Channel Tunnel, opened 1996.

Second Millennium (1800-2000)

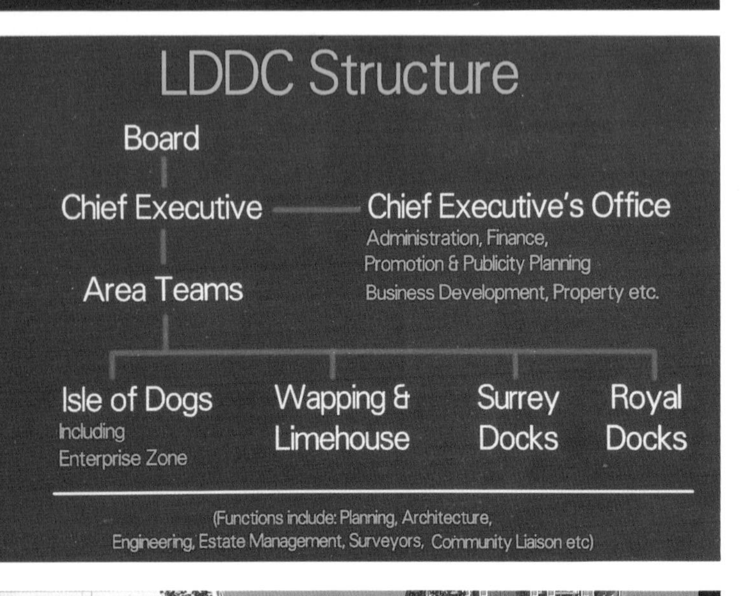

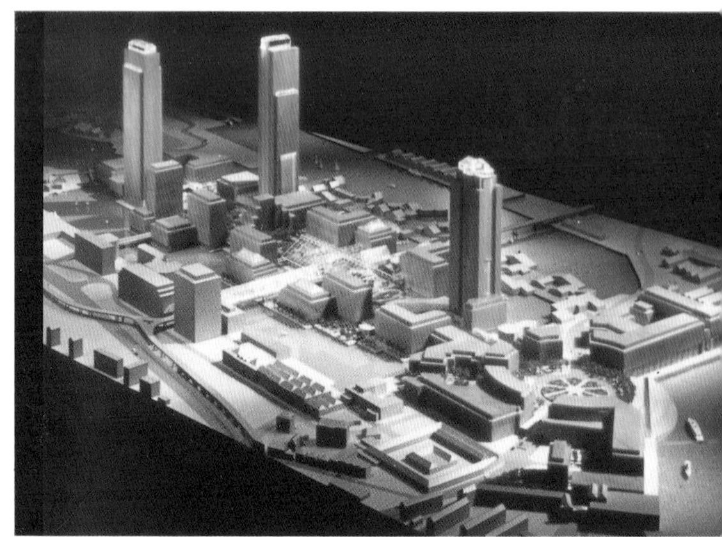

London Docklands Development Corporation

After two centuries as the world's greatest port, London Docklands have been transformed over the past three decades of the twentieth century into a complex of giant skyscrapers, office blocks, elegant homes and riverside apartments, and a major international finance centre. This is recognised as the largest and most successful urban regeneration anywhere and the biggest that has been undertaken in London since the Great fire of 1666.

During the existence of the London Docklands Development Corporation (LDDC) from 1981 to 1988, see charts above, Docklands underwent a remarkable rebuilding. Thousands of new homes have been built, many new companies have moved in and huge commercial floor space has been created.

Docklands with its 88 kilometres of waterfront is one of the most exciting new tourist attractions in London. A wealth of beautiful historic warehouses, old riverside pubs and impressive commercial developments enchant the Londoners and visitors alike. There is much landscaping to admire and picturesque riverside walkways in which to relax and enjoy.

On the Isle of Dogs, Canary Wharf is simply the world's most ambitious urban regeneration project, with its tower being the tallest skyscraper in Europe. The beautiful Westferry Circus, Cabot Square and Canada Square are all complemented by a series of watercourts, terraces and promenades. At Wapping and Bermondsey, the large stock of historic warehouses has created considerable attractions for commercial and residential developments. St Katharine Docks, nestling in the shadow of Tower of London, has become a shrine for Londoners and visitors all year round.

The South Bank and Surrey Docks have been at the centre of amazing transformation with spectacular conversions of Victorian warehouses into luxury apartments and a shopping precinct. The Surrey Docks have a thriving residential community with watersport in Greenland Dock and a yacht marina in South Dock. Wind surfing, sailing, green open spaces, tree-lined walks and nature reserves combine readily with centuries old churches and riverside public houses.

Illustrations

The pictures on the right show the final agreement with the LDDC of Canary Wharf project, a model of the first design and the opening of the Tower in January 1992.

Canary Wharf. The future of London.

Deputy Chairman of West India Dock Co.

The magnificent warehouses on the North Quay shortly after completion 1802.

A panoramic view of West India Docks and warehouses, Isle of Dogs, on the eve of completion 1802 with Canary Wharf at the centre.

Old West India Docks Gateway 1920s.

An artist's impression of CanaryWharf in its heydays of 1950s.

An aerial view of the West India and Millwall Docks in 1967, showing warehouses on the South Quay for discharge of Japanese goods.

Dockers Cottages Isle of Dogs.

A view of Heron Quay and Canary Wharf from the south, circa 1981.

The derelict West India Docks in 1982, showing on the right Canary Wharf sheds formerly used for discharging fruit and vegetables from the Canary Islands.

Round Guardhouse at Canary Wharf.

A view of Canary Wharf warehouses in 1981.

Canary Wharf Tower

The foyer in Canary Wharf Tower.

The dramatic regeneration of Canary Wharf and the Isle of Dogs for a new business centre for the City of London, circa 1995.

REGENERATION OF LONDON DOCKLANDS

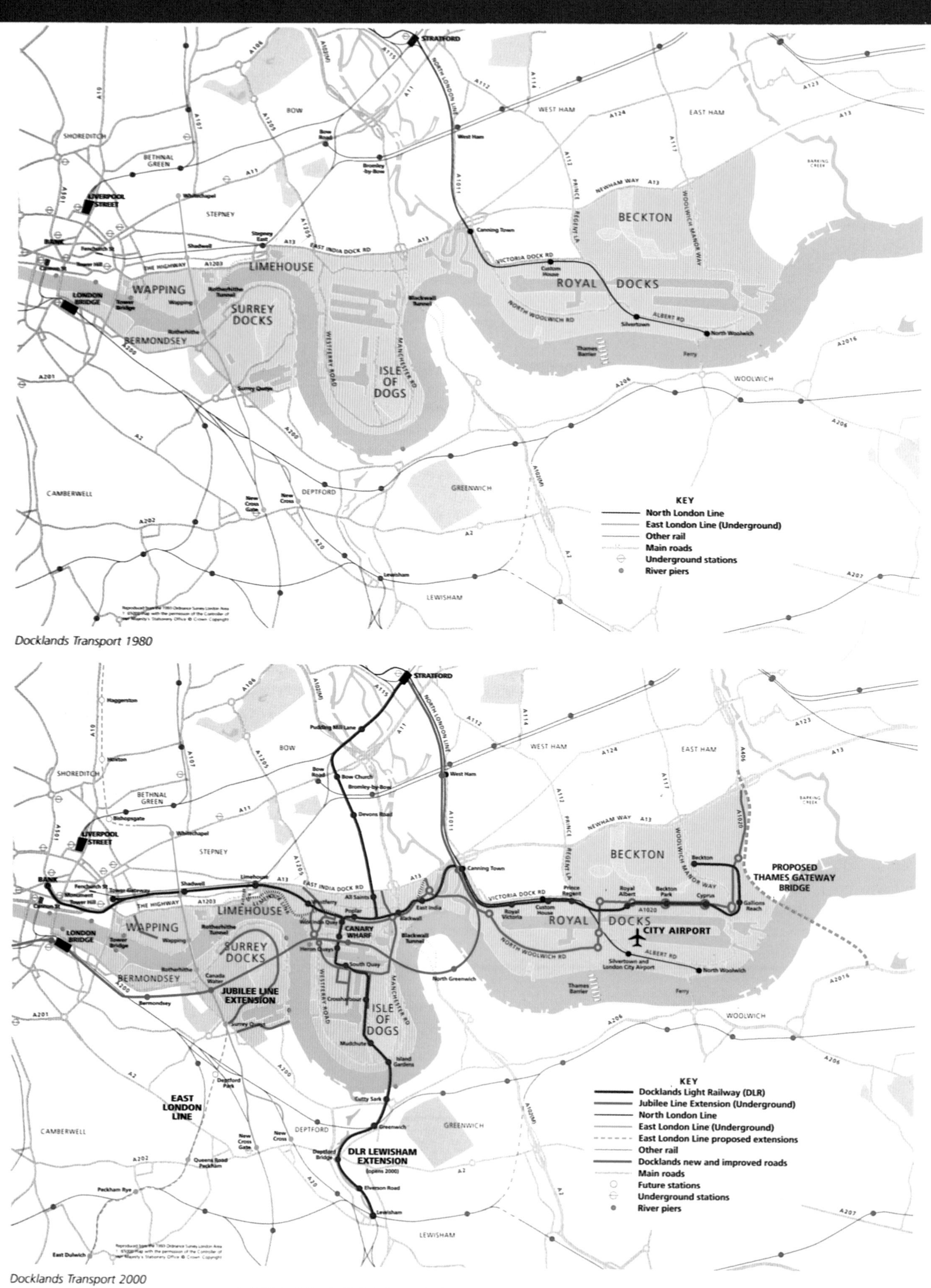

Docklands Transport 1980

Docklands Transport 2000

LONDON DOCKLANDS TRANSPORT NETWORK

London Docklands - Employers

2690

1021

1981 1998

Lifetime Outputs and Cumulative Spend as at 31.3.98:

Population	83,000	(39,400 1981)
Businesses	2,690	(1,021 1981)
Employment	85,000	(27,200 1981)
New dwellings	24,042	
Area of derelict land reclaimed	2,042 acres	
New industrial/commercial floorspace	25.1 million sq ft	
Public sector investment	£1,859 million	
Private sector investment	£7,200 million	

London Docklands Development Corporation (LDDC) 1981 – 1998

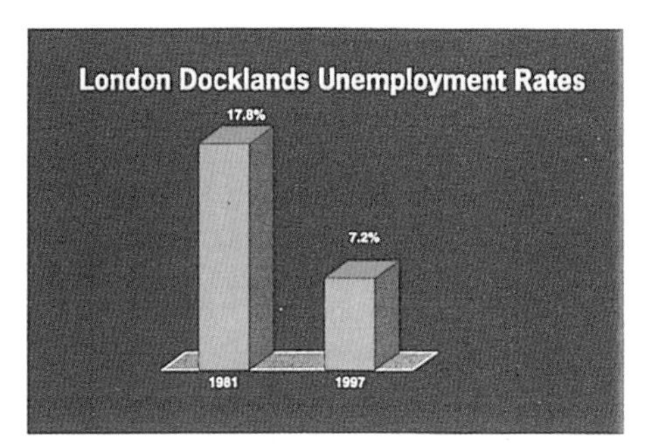

Structure of Local Employment

	Employees 1997	%	Employers 1997	%
Agriculture, Fishing	38	0	6	0
Mining, Quarrying	6	0	2	0
Manufacture	15,765	19	273	10
Energy	94	0	3	0
Construction	2,083	3	112	4
Wholesale, retail, Repair	6,956	8	474	17
Hotels, Restaurants	3,459	4	208	8
Transport, Communication	7,025	9	213	8
Finance Intermediation	17,768	22	123	5
Real Estste, Renting, Business Activities	17,206	21	641	24
Public Administration	2,873	3	35	1
Education	1,744	2	78	3
Health, Social Work	1,986	2	149	6
Community, Service Activities	4,709	6	206	8
Others	4	0	1	0
Unclassified	584	1	166	6
TOTALS	**82,300**	**100**	**2,690**	**100**

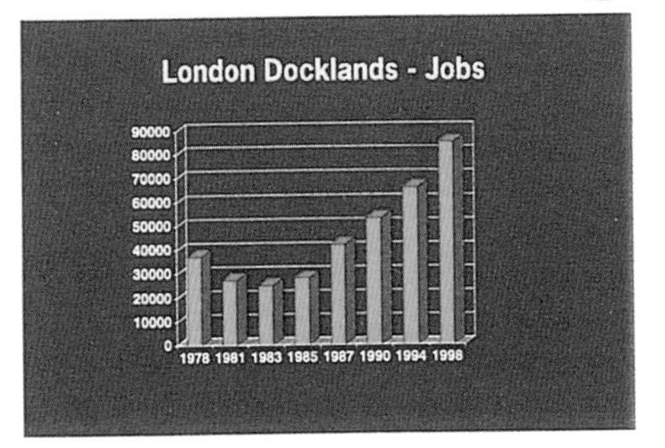

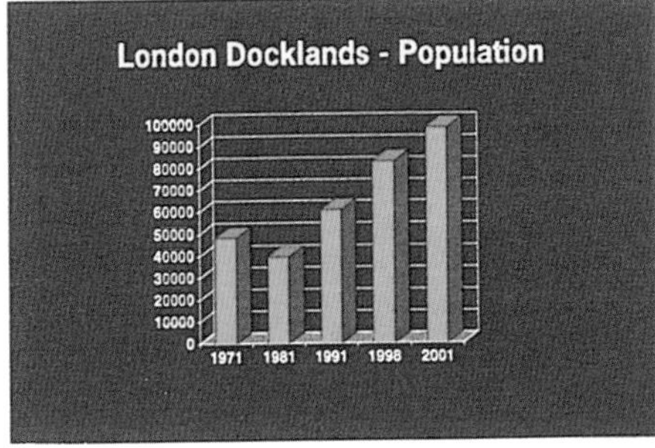

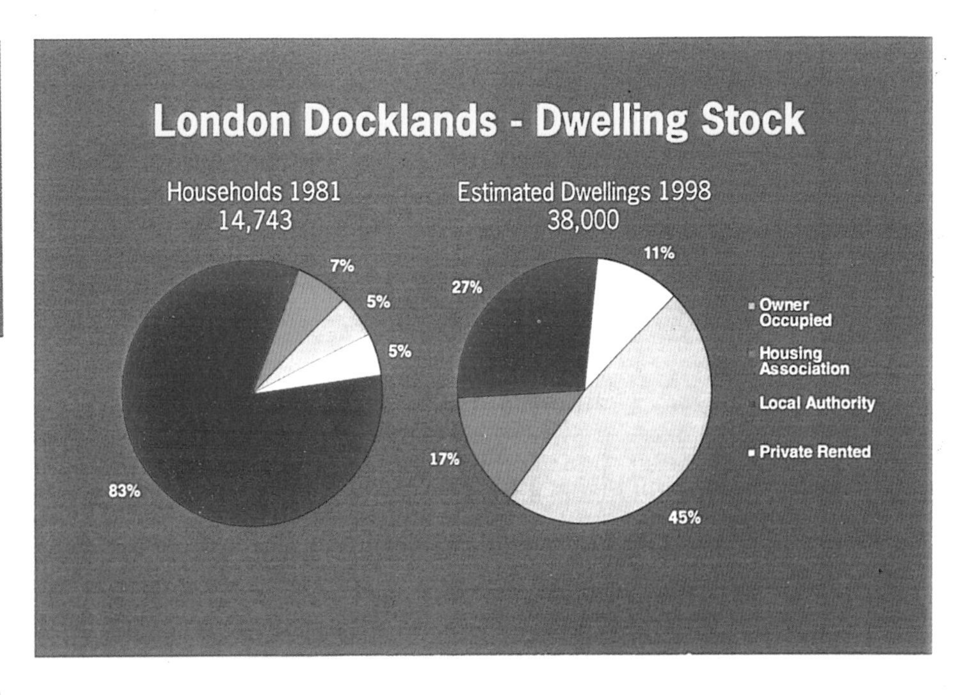

What to see in

City of London,
Crown Jewels & Tower

Built during the reign of William the Conqueror and with the original buildings still remaning, the Tower of London stands as one of the most splendid and significant castles in Europe and houses the Crown Jewels.

Prospect of Whitby Pub At Shadwell

At the entrance to Regent's Canal, Limehouse basin offers fishing and watersports amongst the historic roots of its shipbuilding era.

Canary Wharf Tower

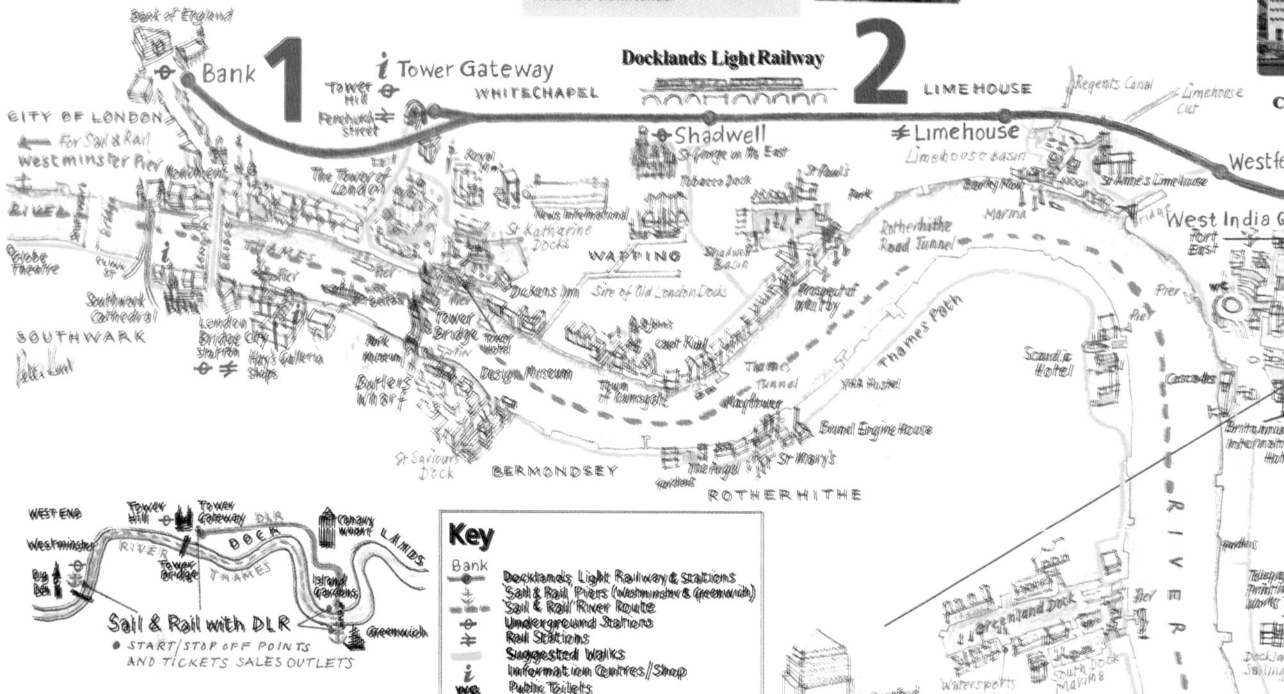

St. Katharine Docks, at the site of the World Trade Centre, offers a pleasant introduction to London's famous docks with its yachts and many interesting shops to visit.

Docklands Light Railway and Jubilee Line interchange is at Canary Wharf.

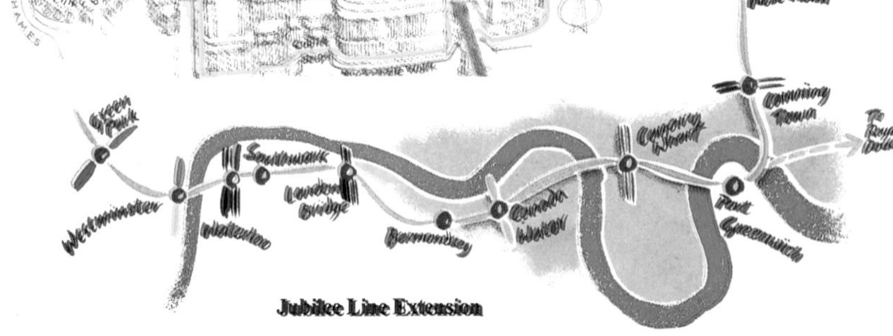

Jubilee Line Extension

Lo
Dock

Area 1 – Bank & Tower Gateway

Bank, the financial heart of the City of London, and Tower Gateway stand at the threshold of Docklands. From the citadel of the Tower of London and Tower Bridge, look West to the City, St. Paul's and the stunning Lloyds of London Insurance building. And look East to the streets of Whitechapel and Aldgate where Jack the Ripper stalked his victims, where John Merryck, the Elephant Man, lived out his life and where the famous Petticoat Lane market still sets out its stalls every Sunday.

Area 2 – Shadwell, Limehouse & Westferry

At the heart of London's East End, this area incorporates the Commercial Road and the Highway, two of London's main historical arteries of trade, linking the City and the Docks. Areas such as Bow, Bethnal Green, Stepney and Wapping are still home to some of London's most celebrated public houses. Enjoy a pint of ale at the oldest riverside inn in the City, the Prospect of Whitby where, over the centuries, Dickens and Samuel Pepys quenched their thirsts.

Area 3 – Stratfo

South of Poplar is the peninsula
Blitz and rebuilt in the late 19
Docklands. The Old Sugar Ware
wharfs where commercial shippi
centres such as the new Billing
development are a testament to

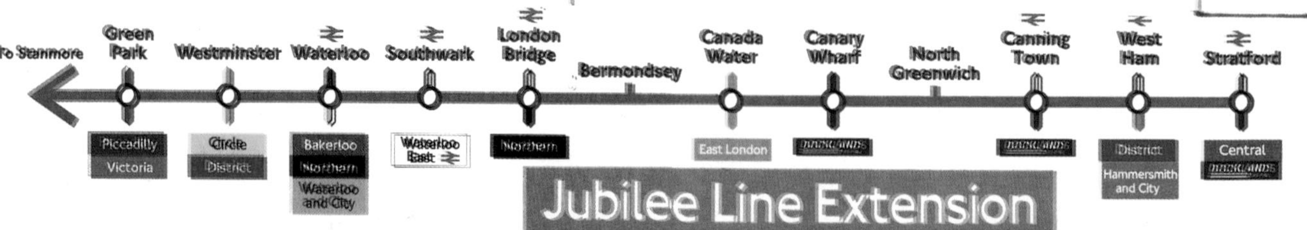

ocklands on DLR

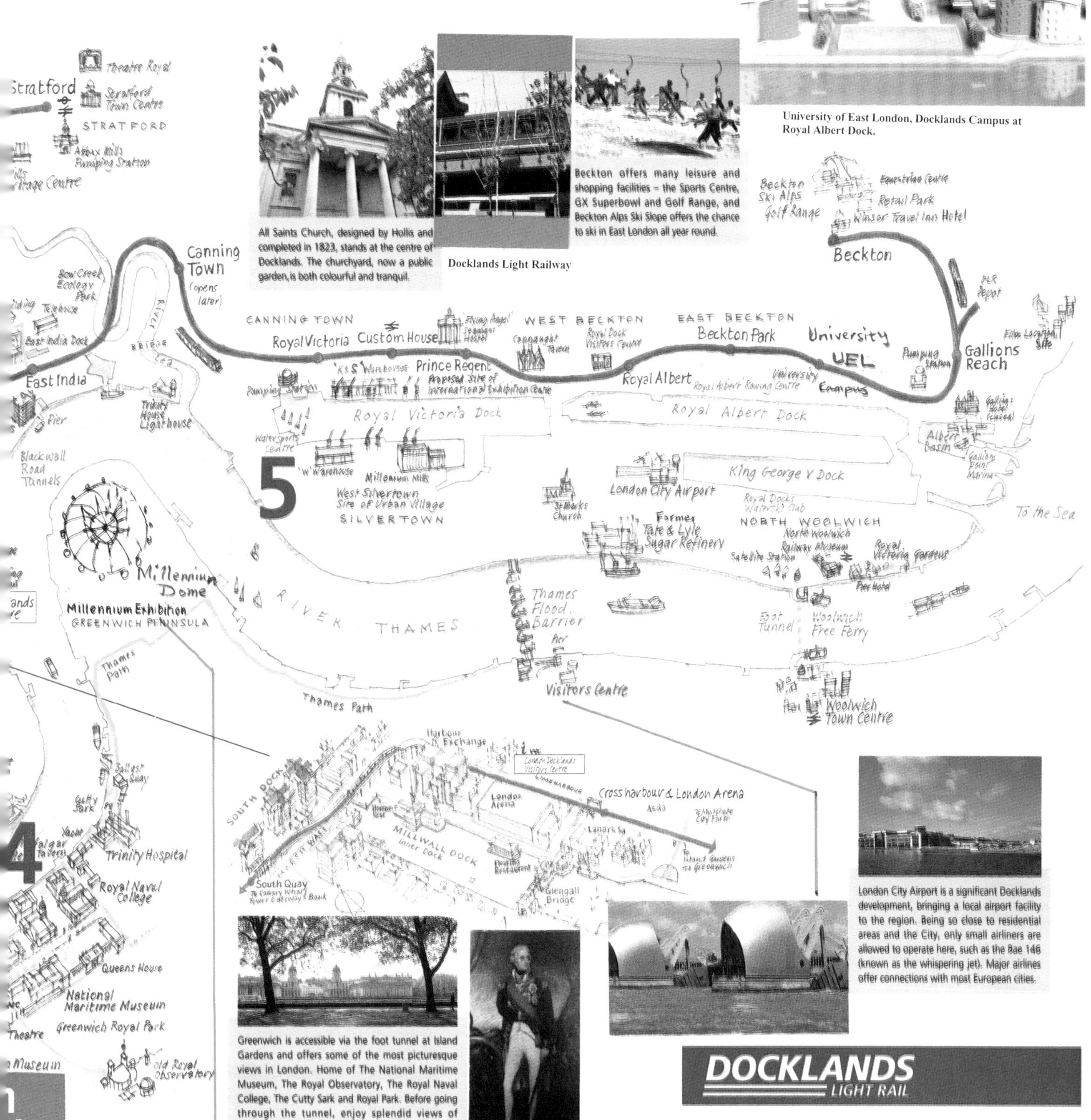

University of East London, Docklands Campus at Royal Albert Dock.

All Saints Church, designed by Hollis and completed in 1823, stands at the centre of Docklands. The churchyard, now a public garden, is both colourful and tranquil.

Docklands Light Railway

Beckton offers many leisure and shopping facilities – the Sports Centre, GX Superbowl and Golf Range, and Beckton Alps Ski Slope offers the chance to ski in East London all year round.

London City Airport is a significant Docklands development, bringing a local airport facility to the region. Being so close to residential areas and the City, only small airliners are allowed to operate here, such as the Bae 146 (known as the whispering jet). Major airlines offer connections with most European cities.

Greenwich is accessible via the foot tunnel at Island Gardens and offers some of the most picturesque views in London. Home of The National Maritime Museum, The Royal Observatory, The Royal Naval College, The Cutty Sark and Royal Park. Before going through the tunnel, enjoy splendid views of Greenwich from the gardens by the tunnel entrance.

...uth Quay

...e of Dogs. Razed to the ground during the ...Dogs is the traditional industrial heart of ... Dockmaster's House sit side by side with ... trade until the 1960s. Modern commercial ...ket and the world famous Canary Wharf ...n of Docklands.

Millennium Dome

Area 4 – South Quay to Greenwich

To truly appreciate the scale of Dockland's resurgence, visit the London Docklands Visitor Centre in Limeharbour. Witness for yourself how the area has been transformed in just a little over ten years. In Coldharbour, you will find a warm welcome in The Gun public house where Admiral Lord Nelson would keep secret romantic meetings with Lady Emma Hamilton. Then, from Island Gardens at the southern limit of the Isle of Dogs, take a walk through the original Victorian foot tunnel under the Thames to Greenwich, home of the National Maritime Museum and the Cutty Sark.

Area 5 – The Royal Docks and Beckton

Now home to London City Airport and the Thames Barrier, historically this area was the mooring point for the largest vessels arriving from the Thames Estuary. For a taste of the country, drop in on Newham City Farm at Stansfeld Road. Gallions Reach, once transformed by Stanley Kubrick into the Vietnam film-set of 'Full Metal Jacket', is still home to The Gallions Hotel, previously used by P&O passengers awaiting embarkation on continental voyages and currently closed for refurbishment. From Royal Albert station, take in some terrific views of London City Airport across a vast expanse of land

The Imperial State Crown.

Mansion House, Lord Mayor's Residence.

The Bank of England in the City.

The Lord Mayor's Show in the City.

Coronation of Her Majesty Queen Elizabeth II, c1953.

Aerial view of the Tower of London, Tower Bridge and River Thames.

The treasures of the Tower of London.

Dickens Inn, pub and restaurant at St. Katharine Docks.

The Yeoman Warders.

The Yeoman Guards at the Tower of London.

St. Katharine Haven with visiting historic ships and barges.

Docklands Railway.

The City, Tower of London and St. Katharine Docks

Old wine vaults showing coopers at work, circa 1888.

Derelict wine vaults at Tobacco Dock, circa 1981.

Conversion of Tobacco Dock vaults into shops, circa 1988.

Gun Wharves converted into apartments next to Wapping Station.

"The Grapes" public house and 18th century ship Captain houses in Narrow Street, Limehouse.

Canary Wharf DLR Station.

Docklands Museum and Sculpture at Canary Wharf.

The pontoon foot bridge at Canary Wharf.

South Quays DLR Station.

East India Dock Office Development.

The entrance foyer at Canary Wharf West.

Westferry Circus at Canary Wharf.

SIGHTSEEING BY DLR WAPPING TO ISLE OF DOGS

Greenwich Millennium Experience

View from Greenwich in the 17th century.

Future prospect of Canary Wharf from Greenwich.

View from Greenwich Park, circa 2000.

The riverside Trafalgar Tavern, built 1830.

Canaletto's painting of the Royal Naval College, circa 17th century.

Foot Tunnel entrance at Island Gardens, opened 1902.

The Old Royal Observatory, built 1658.

The great National Maritime Museum.

Aerial view of the Royal College, Cutty Sark and Queens House.

The Cutty Sark, 19th century sailing clipper.

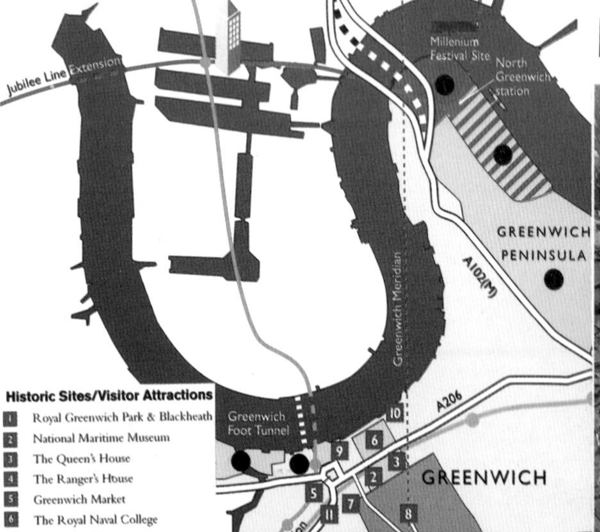

Millennium Dome at Greenwich

Aerial view of Millennium Dome on the Greenwich Peninsula.

Lord Nelson.

Houses of Parliament and Big Ben at Westminster.

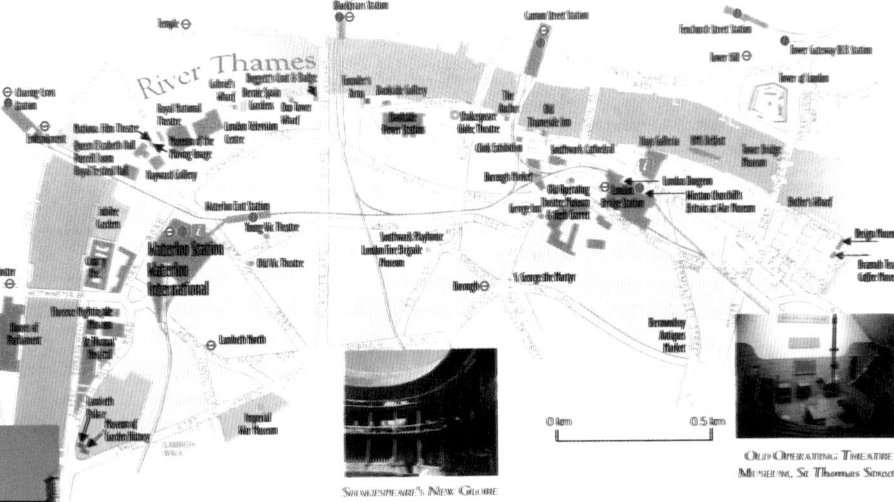

Shakespeare's New Globe

Old Operating Theatre Museum, St Thomas Street.

Map of South Bank.

Former London County Hall.

Aerial View of Royal Festival Hall and South Bank Art Centre.

William Shakespeare.

Opening Ceremony at Globe Theatre.

View from Millennium Wheel.

Millennium Wheel at South Bank.

Shakespeare's Globe Threatre.

London's new footbridge.

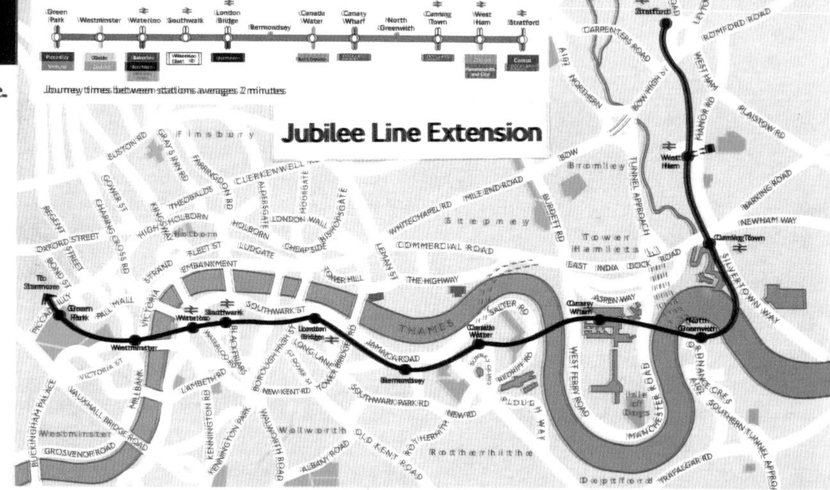

JUBILEE LINE SIGHTSEEING WESTMINSTER TO SOUTHWARK

South of the river is the former London County Hall which has been converted into a hotel with an aquarium in the basement. Walking east you see the Millennium Wheel, followed by the South Bank Art Centre. The complex includes the Royal Festival Hall, Queen Elizabeth Concert Hall, the National Film Theatre and the Hayward Gallery. At Bankside is the new Shakespeare Globe Theatre and the new Tate Gallery. From here a footbridge takes you to the north bank of the Thames and St. Paul's Cathedral.

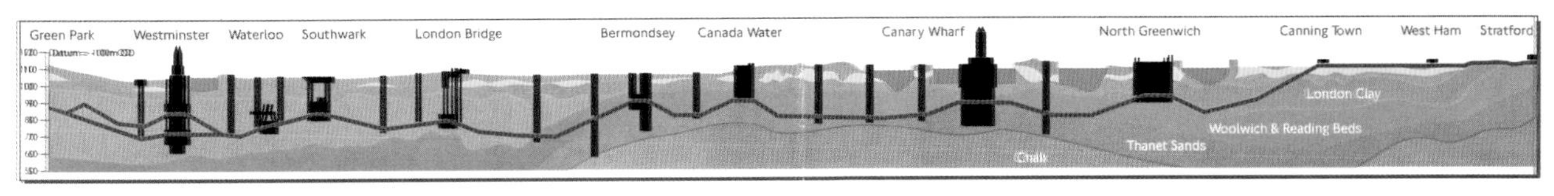

Section through Jubilee Line Extension

Above: Shops and restaurants inside Hayes Galleria.
Below: Panoramic view of Butlers Wharf and Anchor Brew House,

Panorama of East London Docklands on River Thames from Tower Bridge to Canary Wharf.

London Bridge City and Hayes Galleria from the north bank.

HMS Belfast Museum Ship moored at Hayes Galleria and a view of the City of London looking north.

Above: Butlers Wharf and Anchor Brew House today.
Below: Old Butlers Wharf warehouse 1981

Above: Hayes Galleria today.
Below: Hayes Galleria when still a wharf.

JUBILEE LINE SIGHTSEEING LONDON BRIDGE TO BERMONDSEY

On leaving London Bridge station and walking along Tooley Street you find the London Dungeon with its gruesome collection of exhibits from British history. You can also see the exhibition of London during World War II. On the other side of Tooley Street is Hayes Galleria with its shops, restaurants and the new Horniman riverside public house. Moored in the Thames off Hayes Galleria is HMS Belfast Museum Ship which took part in World War II. Walking eastwards you reach Tower Bridge with its museum and elevated walkway which gives excellent views of the river and London. Just down stream of the bridge is the historic area of Shad Thames where a group of waterside Victorian warehouses have been restored including Anchor Brew House, Butlers Wharf and St Saviours Dock. The maze of narrow streets in this area with their tall buildings linked by walkways and bridges forming 'canyons' have been the setting for many period films and dramas. The Design Museum is situated at Butlers Wharf and exhibits the growth of the design process and technology. New Concordia Wharf is an outstanding conversion along St. Saviour's Dock.

London Dungeon museum in Tooley Street.

Above: St. Saviour's Dock apartments today.
Below: Old St. Saviour's Dock warehouses, 1981.

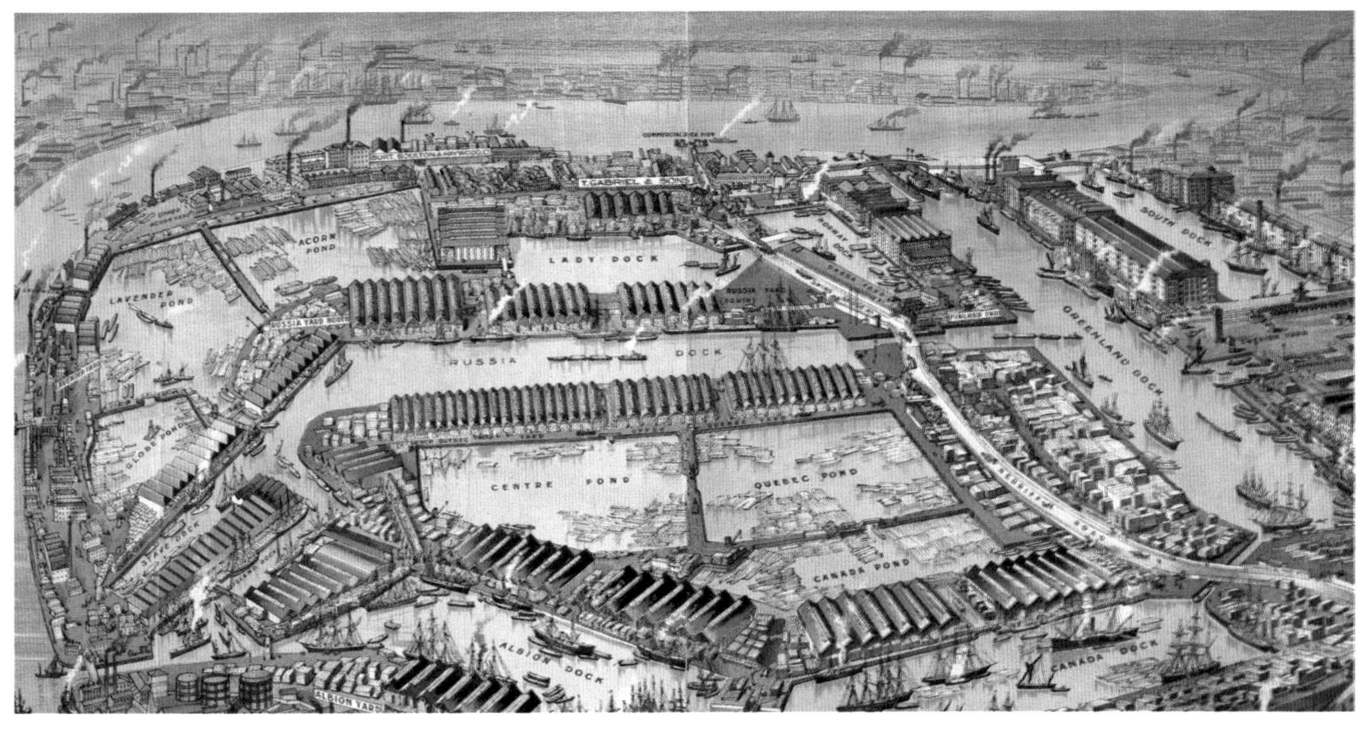

A painting of the Surrey Commercial Docks looking east, circa 1906. They were the centre of timber imports in the UK.

A view towards the City showing the transformation of the Surrey Docks into housing.

Russia Court and Finland Quay housing from Greenland Quay.

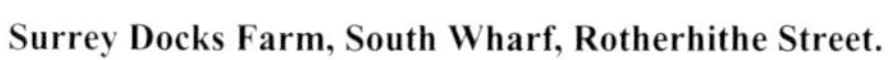

Surrey Docks Farm, South Wharf, Rotherhithe Street.

South Dock yacht marina.

Dr Salter's statue at Cherry Garden Pier.

Tesco Shopping Centre at Surrey Quays.

The Mayflower Pub, Rotherhithe.

The Holiday Inn Hotel, a recent conversion of the old Columbia Wharf, Rotherhithe.

SIGHTSEEING BY JUBILEE LINE SURREY DOCKS AND ROTHERHITHE (CANADA WATER STATION)

London Attractions and Landmarks

London Royal Palaces

Buckingham Palace, in St James Park, originally called Buckingham House, was built between 1702-05 by the first Duke of Buckingham and was bought by King George III in 1762. After John Nash had made considerable enlargements, George IV renamed it Buckingham Palace. The Palace is set in about 40 acres of gardens and houses the Throne Room, Music Room, State Dining Room and Picture Gallery. The Queen entertains many visiting dignitaries at the Palace. Visitors are able to see many of the priceless art treasures, tapestries and furniture as well as the Throne Room.

Windsor Castle is the weekend home of Her Majesty the Queen. You can explore the superb grounds and view the State Apartments which house a collection of treasures including works of art by Van Dyck, Rubens and Rembrandt. During your visit to Windsor Castle you come across what was once an island called Runnymede. You shall see where King John was finally forced to sign the Magna Carta in 1215. There is also the John F Kennedy Memorial on Meadows in the same area. In the grounds of Windsor Castle you can view the changing of the Guard. You can also visit St George's Chapel where you can see a great wealth of treasures and also the tombs of King Henry VIII and Charles I.

Hampton Court lies on the River Thames and is a perfect setting for the famous Tudor Palace built in 1514 by Cardinal Wolsey who gave the Palace to King Henry VIII. It is the largest brick built residence in the United Kingdom and has marvellous traditional gardens laid out in the 17th century.

Whitehall and Houses of Parliament

You walk along Whitehall past Downing Street, the official home of the Prime Minister, the Cenotaph, Trafalgar Square with Nelson's Column, to Buckingham Palace, the London home of the Queen, to see the Changing of the Guard. Walk back to Trafalgar Square, convenient for shopping or further sightseeing. There are many panoramic coach tours in which you can see the highlights of the City of London and the West End, enlivened with interesting sights and facts described in this book. Here you can see Big Ben, the Houses of Parliament, Westminster Abbey, Downing Street, Buckingham Palace, Trafalgar Square and Piccadilly Circus, in this area of magnificent buildings, delightful squares, beautiful parks and smart shopping streets.

London West End

In the West End of London you can see some of the major shopping and entertainment centres, along with some of the Royal Parks. In elegant Mayfair and South Kensington you see arcades and the world famous Harrods and some of the most renowned museums. In South Kensington you can also visit Kensington Palace, St James' Palace and Albert Hall. The National Gallery is a famous landmark in Trafalgar Square. In Convent Garden you can stroll through this traditional centre of activity with street entertainers, old-world pubs and lively restaurants with a chance to jostle for some souvenirs in the small shops. You can also visit the museum of London Transport, the Royal Opera House and the Church of My Fair Lady. At Buckingham Palace you will see the Changing of the Guard and make your way to the Queen's Gallery. Nearby in Whitehall you can also see the Guards Change at Horseguards. Your trip to London is incomplete without experiencing a traditional pub which are dotted all over the centre of London, some of which are acclaimed Victorian Inns and waterside pubs. Relax in the unique British atmosphere and taste some of the excellent beers. A fitting end to a tour is to cruise say from Tower Bridge to Festival Pier while enjoying the traditional past time, tea with biscuits or scones.

The City of London

From the West End of London you can travel to the historic City of London. The City dates back to Roman times and is full of fine historic buildings. Here you will see the Stock Exchange, Bank of England, the Old Bailey and many others before you arrive at St Paul's Cathedral. This was built by Sir Christopher Wren in 1710 and is not only the final resting place of renowned soldiers, statesmen, painters and poets, but also scene of many famous weddings – most notably that of Prince Charles and the late Lady Diana.

The Tower and Crown Jewels

On the boundaries of the City is the Tower of London, a fortress dating back over 900 years. The Tower was used as a fortress, prison, palace and place of execution. The atmosphere within the Tower is immersed with haunting memories of torture and imprisonment as well as the richness of royal ceremonies it has hosted. Here you can see the magnificent crown jewels, meet the Beefeaters in their traditional Tudor uniform and view one of the world's greatest display of medieval armour. Steeped in the nation's history, it has been the scene of many executions, including the be-heading of Henry VIII's wife, Anne Boleyn.

Bus and Coach Tours

Panoramic tours are organised by a number of bus and coach operators. You can join the circular tour at any one of their pick-up points and see London from the top of an open bus. You will see the major sites in London including the Tower of London, St Paul's Cathedral, Whitehall, Trafalgar Square and Piccadilly Circus. You can listen to a pre-recorded commentary in English which will tell you about the well known buildings and landmarks en route. The bus will also cross the River Thames a number of times and you can see Westminster from different vantage points.

London's Sunday Markets

Every Sunday some of London's main streets give way to stalls and become street markets for the day. People from all parts of the City join for an unusual morning of shopping and walking as one of their favourite past times including rummaging through the stalls as these friendly but noisy street markets appear to the visitors. Petticoat Lane is a market which has been held for over 300 years. You can buy and sell literally anything, sometimes you may be lucky and find a bargain. At Spittlefields new venue where once stood a good fruit and vegetable market, here you can wonder around under cover and search for that item you really need. The Greenwich Craft Market is an area of narrow streets, quaint shops and wonderful stalls. If you are tired of shopping, visit the famous sailing ship, the Cutty Sark, or walk up the hill to the Observatory. Remember you are on the Meridian GMT. There are many other interesting Sunday markets all over Greater London.

Thames Cruises

Enjoy the unique setting of the historic River Thames to relax. Catamaran cruises have bar and restaurant facilities and offer commentary pointing out the places of interest which can be seen from either the inside or outside viewing areas. Cruises are normally arranged from Westminster Pier to Greenwich and the Thames Barrier, and westwards to Hampton Court. Catch a sightseeing cruise from any of the five piers and relax on your way to and from Charring Cross, Tower of London, St Katharines, Greenwich and the Thames Barrier while enjoying the magnificent views along the Thames. You will see London's most famous sites, Big Ben and the Houses of Parliament, St Paul's Cathedral and the City, Tower of London, Tower Bridge to name a few. You can also have a night tour of cruising with a dinner dance afloat with live music, it is a different way of seeing London at night with panoramic windows and exterior floodlights.

Maritime Greenwich

After joining the cruiser at Charring Cross Pier you will enjoy a cruise downriver to Greenwich. You will see many famous landmarks including HMS Belfast, Tower Bridge, the Tower of London and Canary Wharf Tower – Europe's tallest building. The Captain will point out the major landmarks as you pass by. At Greenwich you visit the Royal Observatory situated on the 'O' meridian which can be seen on top of Greenwich Hill. Its descending moon time bowl was a time signal to Londoners for many centuries. You can also visit the world renowned Maritime Museum and Queen's House.

The Museum Experience

Great Britain has more museums to the square kilometre than any other country in the world. Most of these (over 1600) are located in London and housed in cultural buildings with classical façades, others in disused homes and windmills. The greatest have paintings, ancient treasures, furniture, porcelain, metalwork and jewellery worth £billions. Some are magpie collections or hoards by speculative specialists who amassed dolls, pianos, stuffed animals or fancy dress or product design.

Exceptional collections include some of the world's artworks of human civilisation. Super items include the Elgin Marbles, the Alfred Jewel, the Portland Vase (smashed by a lunatic in 1845 but brilliantly restored by the British Museum, and the Great Bed of Ware in the Victoria and Albert Museum at South Kensington, a piece of antique mentioned by Shakespeare. Small museums contain items of history such as a watchcase enamelled with heartcase flowers, an 18th century barber/surgeon English/Delft bleeding bowl and Jacobean silk stockings. There are many more categories of museum, for example children, military, folk, science, transport etc.

London Entertainment

Sensational London entertainment can be found in three famous places. The Beefeater in Ivory House, St Katharines Dock has a traditional Medieval banquet with an action packed evening of entertainment by jesters, minstrels and 'Royal entertainers'. At the London Cockney in Tottenham Court Road, you can experience an evening of traditional London hospitality with variety music hall entertainment with unique cabaret and live band. At the Talk of London in Drury Lane, Parker Street, you can enjoy an evening of dining in elegant surroundings including a cabaret from Talk of London dancers and featuring a resident band. There are also many theatres in the West End with world famous shows which must not be missed.

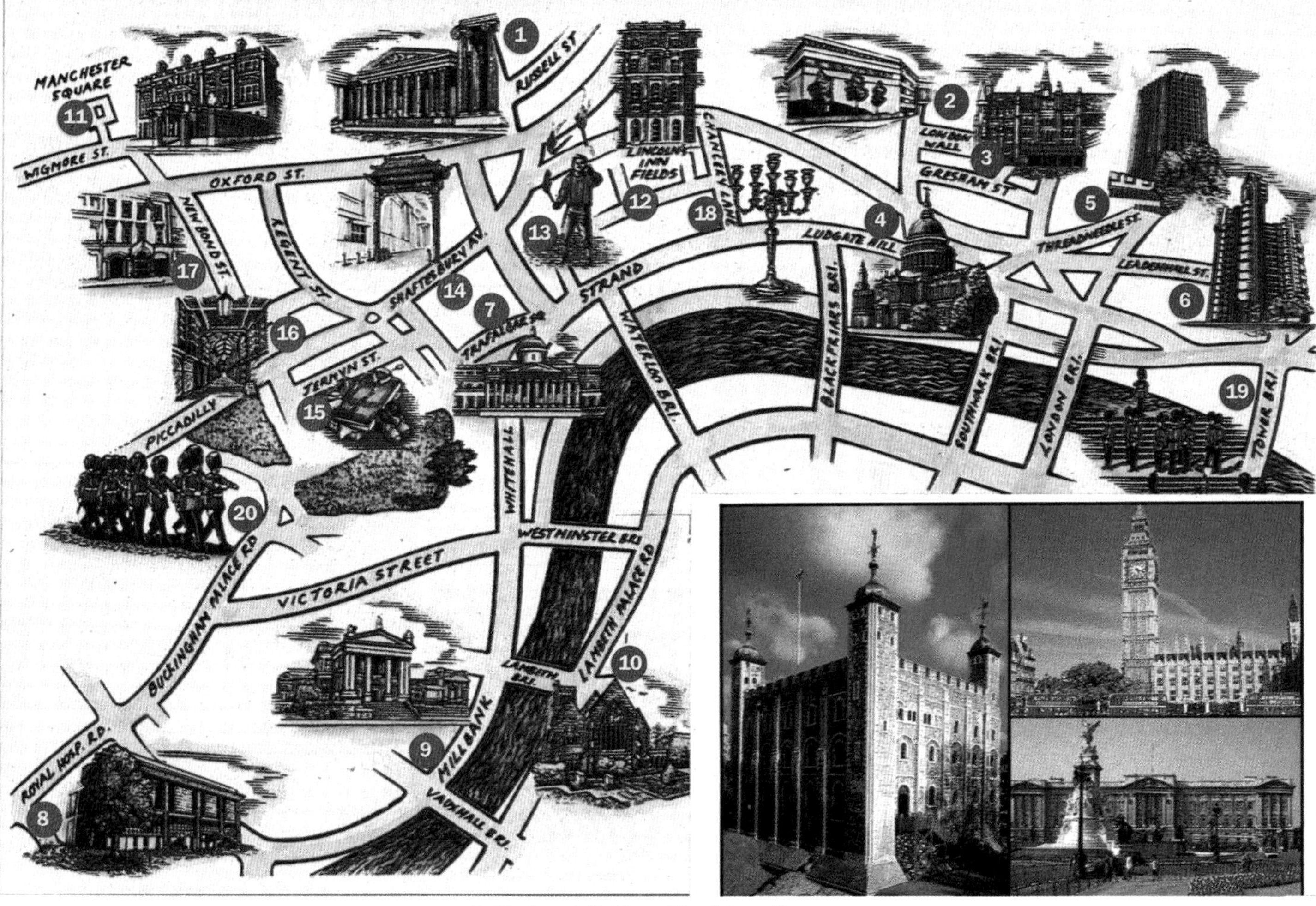

Selected London Attractions

1. British Museum, Bloomsbury

After 240 years, this is still the biggest attraction in London. Its vast store of treasures range from the Elgin Marbles to 'Ginger', the 5,000 year-old mummified Egyptian.

2. Museum of London, Barbican

This is the best place to understand how London grew. Look for the Roman bikini, the Great Fire Experience of 1666 and an unexploded bomb from the Second World War.

3. Guildhall, City of London

This is the medieval seat of London's power. The Lord Mayor, in full regalia, presides over the monthly Court of Common Council in the Great Hall, accompanied by a Sword Bearer, Mace Bearer and City Marshall.

4. St Paul's Cathedral, City of London

To appreciate the soaring sound of the choir and organ filling the second largest church dome in the world, go to Evensong (5.00pm Monday-Saturday; 3.15pm Sunday). Ask a verger if there's room to sit in the Grinling Gibbons choirstalls.

5. Stock Exchange, City of London

Book in advance for a lecture, video and sight of the old trading floor, where colourful computer screens record the ups and downs of the FT index.

6. Lloyds, City of London

Prince Charles may not like the architecture but visitors are awe-struck by the modern building, with its bird's eye view of the underwriting floor and the Lutine Bell. There is stunning lighting at night, thanks to theatrical illuminations.

7. National Gallery, Trafalgar Square

Over 2,000 great paintings to appreciate. Take a guided tour or a lunchtime lecture. A pamphlet lists the best-known masterpieces.

8. National Army Museum, Victoria

You can see the skeleton of Napoleon's horse, Wellington's telescope from Waterloo and Florence Nightingale's lamp, plus lots of uniforms.

9. Tate Gallery, Embankment

The new Clore Gallery housing the Turner Bequest is worth a detour; then there's the modern collection, as well as the new exhibition at Bankside.

10. Museum of Garden History

The botanists and gardeners royal, are buried within St. Mary-at-Lambeth. Visit the 17th century-style garden and Lambeth Palace.

11. Wallace Collection, West End

This is best known for *The Laughing Cavalier* but with exquisite French furniture and porcelain as well.

12. Sir John Soane Museum, Holborn

London's best-kept museum secret: this personal collection remains as it was when Sir John was alive, two centuries ago. Ask for the panels in the Picture Room to be opened.

13. Covent Garden, West End

This is a bustling area of central London every day. Applaud the art of the buskers in front of St Paul's Church and the glassblowers at the Glasshouse in Long Acre.

14. Chinatown, Soho

The pagoda-like telephone booths and Chinese street signs may look interesting but the many Chinese diners confirm the wind-dried duck is the real thing.

15. Jermyn Street, West End

Window-shop at Royal Warrant Holders like Paxton and Whitfield (cheese) and Floris (perfume). Prince Charles gets his shirts from Turnball and Asser.

16. Burlington Arcade, West End

London's oldest shopping mall. Don't whistle, run or open your brolly - the Dickensian beadle's watching you!

17. Sotheby's, West End

See what's 'on view' before it goes under the hammer at the world's oldest art auctioneers. Could be a million-pound painting or a £200 vase. You may bid for Elvis's jump-suit, a Marilyn Monroe dress or Jimi Hendrix guitar!

18. London Silver Vaults, City

130 vaults store the world's largest collection of silver 30 feet below Chancery Lane. The dealers hope you will be dazzled.

19. Ceremony of the Keys, Tower of London, City of London

At 700 years, the longest running show in town. Book up three months ahead to experience the dark power of the Tower. Apply in writing only, with a choice of dates and enclosing an SAE to: Resident Governor, HM Tower of London, London EC3N 4AB.

20. Changing of the Guard, Buckingham Palace, West End

For a front row view of the pomp and ceremony, be outside Buckingham Palace at 10.30am, an hour ahead of time.

St Georges Chapel at Windsor Castle

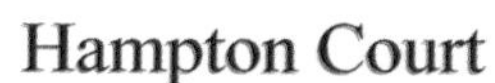

~ History is bought to life by the costumed guides in the KING'S APARTMENTS.

~ The stately EAST FRONT of the Palace seen from the GREAT FOUNTAIN GARDEN.

~ The POND GARDENS, just one of many features making up 70 acres of riverside gardens.

~ Enter WILLIAM III'S PRIVATE STUDY where he signed letters and attended to state business.

~ The ASTRONOMICAL CLOCK made for HENRY VIII in 1540.

~ Feel the heat of the roaring log fire and the boiling cauldrons in the cavernous TUDOR KITCHENS.

~ Dating from HENRY VIII's time, the CHAPEL ROYAL is a stunning example of the Palace's rich interiors.

~ The statue of the THREE GRACES was cast in Paris and dates from the early nineteenth century.

Experience the Treasures of Royal London

Piccadilly Circus and Statue of Eros.

Aerial views of Trafalgar Square and Picca

Trooping the Colour, Horse Guards Parade.

Westminster Bridge, Houses of Parliament and Big Ben.

A view along Whitehall from Westminster

cus.

Horse Guard.

Trafalgar square and the National Gallery.

Carnaby Street and Harrods of Knightsbridge.

Selfridges of Oxford Street and Liberty of Regent Street.

London West End Landmarks

lgar Square.

L O N D O N G

1 Barbican Art Gallery

Level 8, Barbican Centre, London EC2Y 8DS
☎ 071 638 4141 X7619 Groups 071 638 8891

Opening hours: Mon, Weds-Sat 10:00-18:45, Tues 10:00-17:45
Entry charge

The Barbican Centre, Europe's largest Arts Centre and home to the Royal Shakespeare Company and the London Symphony Orchestra, contains two art galleries. Barbican Art Gallery is a major venue for changing exhibitions and the largest exhibition space in the Barbican Centre, offering a varied programme of exciting and innovative shows concentrating mainly on 19th and 20th-century painting, sculpture and photography. The sculpture of Eric Gill, the 60s art scene in London, Alphons Mucha and the photography of Bill Brandt are some of the features for 1993. The Concourse Gallery offers an unconventional focus for some of the best free shows in London.

Eric Gill, *Splits II*, (front view), 1923 (Art Collection, Harry Ransom Humanities Research Centre, The University of Texas)

Robyn Denny, *Austin Reed Mural*, 1959

⊖ *Barbican or Moorgate* ✕ ♿ P 🚌

2 Hayward Gallery

Belvedere Road, The South Bank Centre, London SE1 8XZ
Recorded Information: 071 261 0127 Advance Bookings: 071 928 8800
Group Bookings: 071 921 0849

Opening hours: Thurs-Mon 10:00-18:00, Tues & Weds 10:00-20:00.
Closed between exhibitions. Entry charge depends on exhibition.

The Hayward Gallery has been the originator or host of many of the world's most influential exhibitions since 1968. It mounts temporary exhibitions of contemporary and historical art including Magnum Photographers, Chinese Paintings, Twilight of the Tsars, Doubletake: Collective Memory and Current Art, The Art of Ancient Mexico as well as exhibitions devoted to the work of individual artists - Leonardo da Vinci, Andy Warhol, Jasper Johns, Richard Long, Toulouse-Lautrec and Magritte. A Gallery Shop selling catalogues and posters for past and present exhibitions, postcards and art books is open to the public during exhibitions. An education programme accompanies all exhibitions.

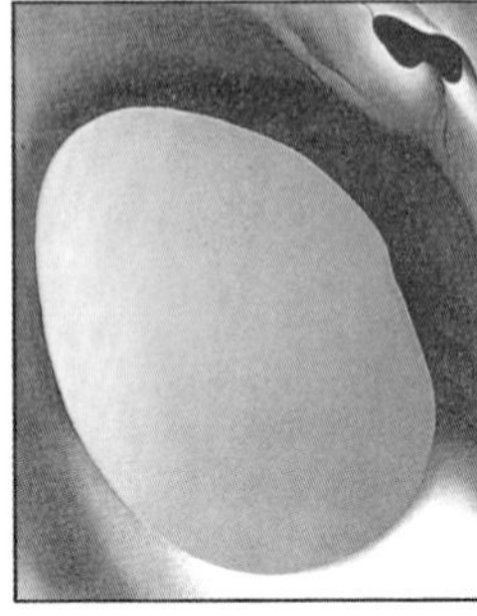

Georgia O'Keeffe, *Pelvis series*

Guilio Paolini, *Hi-Fi* (Gravity and Grace Exhibition)

⊖ *Waterloo or Embankment*  ✕ ♿ ⇌ P 🚌

3 Royal Acade

Burlington House, Piccadilly, Lond
☎ 071 439 7438 Groups 071 494

Opening hours: Mon-Sun 10:00-18: admission 17:30)
Entry charge varies with exhibition.

The Royal Academy of A famous for its varied and exciting programme of ir national art exhibitions. exhibitions, held in some Europe's most beautiful g leries, including the new award-winning Sackler Galleries, cover all aspec art. Recent blockbusters have included 'Monet in 90s', 'The Pop Art Show' 'Andrea Mantegna'. The Royal Academy also has extensive permanent coll tion which includes pain sculpture, prints and dra ings, the highlight of whi

Edward Hopper, *From Willi*

6 Courtauld Institute Galleries

Somerset House, Strand, London WC2R ORN
☎ 071 873 2526 Groups 071 873 2549

Opening hours: Mon-Sat 10:00-18:00 Sun 14:00-18:00.
Entry charge

Somerset House is a premier neo-classical building in London (1776-80) which now houses the internationally famous Samuel Courtauld Collection of Impressionist and Post Impressionist paintings and the Princes Gate Collection of Old Master Paintings: also Gambier-Parry, Lee, Fry and Hunter Collections. Temporary exhibitions in the Prints and Drawings Exhibition Room.

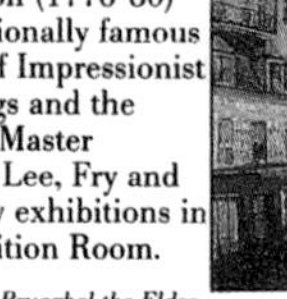

Peter Paul Rubens, *The family of Jan Brueghel the Elder*

⊖ *Temple or Covent Garden* 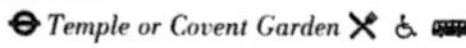 ✕ ♿ 🚌

8 National Gallery

Trafalgar Square, London WC2N 5DN
☎ 071 839 3321 Groups 071 389 1744

Opening hours: Mon-Sat 10:00-18:00 Sun 14:00-18:00.
Entry charge for special exhibitions only

The National Gallery houses the permanent National Collection of Western European Painting from c.1260-1920. The collection is divided into four parts and located in four different areas of the building: Sainsbury Wing - painting from 1260-1510, West Wing - 1510-1600, North Wing - 1600-1700, East Wing - 1700-1920.

'The Wilton Diptych', c. 1395 (The National Gallery, London)

⊖ *Leicester Square, Piccadilly Circus or Charing Cross* ✕ ♿ ⇌ 🚌

7 Design Museum

Butlers Wharf, Tower Bridge, Shad Thames, London SE1 2YD
☎ 071 403 6933 Groups 071 407 6261 (Recorded information)

Opening hours: Mon-Sun 10:30-17:30.
Open Bank Holidays. Entry charge

The Design Museum is the first museum in the world to examine the role of design in our everyday lives. Changing displays include furniture, domestic appliances, cars, graphics and ceramics, and exhibits have been as varied as the Citroën DS, the Mae West Lip Sofa and the prototype of the 1992 Olympic gold medal-winning bicycle.

⊖ *Tower Hill or London Bridge* ✕ ♿

9 National Maritime Museum & Queen's House

Romney Road, Greenwich, London SE10 9NF
☎ 081 858 4422

Opening hours: Apr-Sept 10:00-18:00 Sun 12:00-1800. Oct-Mar 10:00-17:00 Sun 14:00-17:00. Entry charge

Home of the world's most important collection of maritime art and artefacts. From holdings of 3,000 oil paintings including works by Turner, Hogarth, Canaletto and Reynolds, several hundred are on permanent display in these impressive historic buildings. Also clocks, sundials, 17th century ship models, plate, swords and gilded Royal barges.

Van de Velde, the Younger, *HMS Resolution in a gale*, 1633-1707

✕ ♿

Rubens's ma
the Early Mo

From Turner, Constable, Brueghel and Rembrandt to Warhol, Hockney, Bacon and Moore - here is your guide to some of the great riches of fine and decorative art in both permanent collections and temporary exhibitions

Start your journey in London: we list thirteen of its finest galleries. Pissarro comes to the Royal Academy of Arts from July to October 1993 and new at the Victoria and Albert Museum are the Frank Lloyd Wright and 20th Century Galleries. From March to June the Barbican Art Gallery looks at the Sixties in London -Riley, Hockney, Ayres and others. There are major retrospectives of the pioneering American modernist Georgia O'Keeffe at the Hayward Gallery on the South Bank of the Thames and of Ben Nicholson at the Tate Gallery.

Discover the regional galleries of England, from the award-winning modern architecture of the Sainsbury Centre in Norwich to an extravagant Victorian villa in Bournemouth, from Pre-Raphaelites in Manchester to contemporary sculpture in the open air in Yorkshire.

Scotland's beautiful capital, Edinburgh, houses three outstanding Scottish national collections, while Glasgow's museums include the world famous Burrell Collection.

The National Museums of Wales offer not only the stunning collection in Cardiff but also the fascinating Graham Sutherland Gallery in Haverfordwest.

Entry to many galleries is free - so enjoy your journey around Britain and discover Great British Galleries.

Map of London Galleries

4 Tate Gallery

Millbank, London SW1P 4RG ☎ *(071) 821 1313*

Opening hours: Mon-Sat 10:00-17:50 Sun 14:00-17:50.
No entry charge (except for major exhibitions)

Founded in 1897, the Tate Gallery houses the national collection of British painting from the 16th century to the present day, the international collection of 20th century paintings and sculpture and the Turner collection in the specially designed Clore Gallery. Displays change annually, supplemented by a full programme of special loan exhibitions, free lectures, films and talks. The Tate Gallery 'New Displays 1993' will include work by artists such as Hogarth, Gainsborough and Constable from the British collection and Picasso, Matisse, Cézanne, Rodin, Modigliani, Moore, Bacon, Rothko and DeKooning from the modern collection.

Henri Matisse, *The Snail*, 1953 (The Tate Gallery)

JMW Turner, *Norham Castle, Sunrise*, c.1845

⊖ *Pimlico*

5 Victoria and Albert Museum

Cromwell Road, South Kensington, London SW7 2RL
☎ *071 938 8500*

Opening hours: Tues-Sun 10:00-17:50 Mon 12:00-17:50. Entry by voluntary donation.

The Victoria and Albert Museum is the world's finest museum of the decorative arts. The Museum, founded in 1851 by Prince Albert, husband of Queen Victoria, houses many of the world's greatest art treasures. Here you will find the national collections of Sculpture, Furniture, Textiles and Dress, Ceramics and Glass, Silver, Jewellery and Metalwork drawn from European and non-Western cultures.

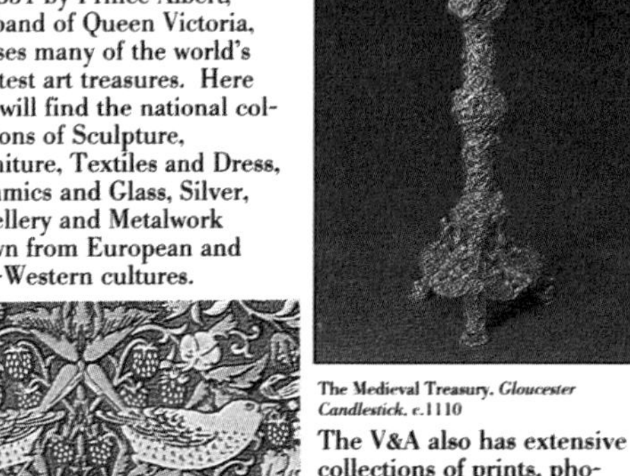

The Medieval Treasury, *Gloucester Candlestick*, c.1110

William Morris, Printed Textile: *Strawberry Thief*, 1883.

The V&A also has extensive collections of prints, photographs and paintings, including the Constable collection. New galleries include the 20th Century Gallery, Designs of the Times, Frank Lloyd Wright Gallery and the Samsung Gallery of Korean Art. There is a changing programme of exciting temporary exhibitions.

⊖ *South Kensington*

…rts

…elangelo's marble relief …e 'Virgin and Child with …nfant St John'.

…ngelo, *The Taddei Tondo: The Virgin …d with the Infant St John* 1504-1505

…iccadilly or Green Park

…ce, An Autumn Landscape with a View of Het Steen in
…'s one of the greatest treasures of the National Gallery

10 National Portrait Gallery

St Martin's Place, London WC2H OHE ☎ *071 306 0055*

Opening hours: Mon-Fri 10:00-17:00 Sat 10:00-18:00 Sun 14:00-18:00. No entry charge

The National Portrait Gallery was founded in 1856 to collect the likenesses of famous British men and women. The collection is the most comprehensive in the world showing art from the Tudors to the present day. Temporary exhibitions throughout the year, plus a regular programme of free lectures, films and workshops.

G.F.Watts, *Ellen Terry*, 1864

⊖ *Charing Cross or Leicester Square*

11 The Queen's Gallery

Buckingham Palace, London SW1A 1AA
☎ *071 799 2331 Groups 071 930 4832 X3351*

Opening hours: Tues-Sat 10:00-17:00 Sun 14:00-17:00 (Closed Mons except Bank Holidays). Entry charge

The Queen's Gallery was opened in 1962 to hold exhibitions based on the Royal Collection, one of the finest art collections in the world. The exhibition planned for 1993 is entitled 'A King's Purchase: George III and the Collection of Consul Smith'.

Queens Gallery

⊖ *Victoria, St James's Park or Green Park*

12 Wallace Collection

Hertford House, Manchester Square, London W1M 6BN ☎ *071 935 0687*

Opening hours: Mon-Sat 10:00-17:00 Sun 14:00-17:00. No entry charge

This permanent exhibition of superb paintings, 18th century French furniture, Sèvres porcelain, clocks and objets d'art, plus one of the finest array of arms and armour in the world was bequeathed to the nation by Lady Wallace in the late 19th century. It is still displayed in the luxurious house which belonged to the collectors.

Sir George Clausen, *The Boy and the Man*, 1908

⊖ *Bond Street or Baker Street*

13 Whitechapel Art Gallery

80-82 Whitechapel High Street, London E1 7QX
☎ *071 377 0107*

Opening hours: Tues-Sun 11:00-17:00 Wed 11:00-20:00 Closed Mon. No entry charge (except for one exhibition per year)

The Whitechapel opened in its purpose-built gallery in 1901 and has an international reputation as one of the key galleries in Europe showing modern and contemporary art. Exhibitions include reappraisals of major 20th-century figures and introductions to important contemporary artists. Free talks and films. Café and bookshop.

Juan Gris, *Newspaper and Fruit Dish*, 1916 (Yale University Art Gallery)

⊖ *Aldgate East*

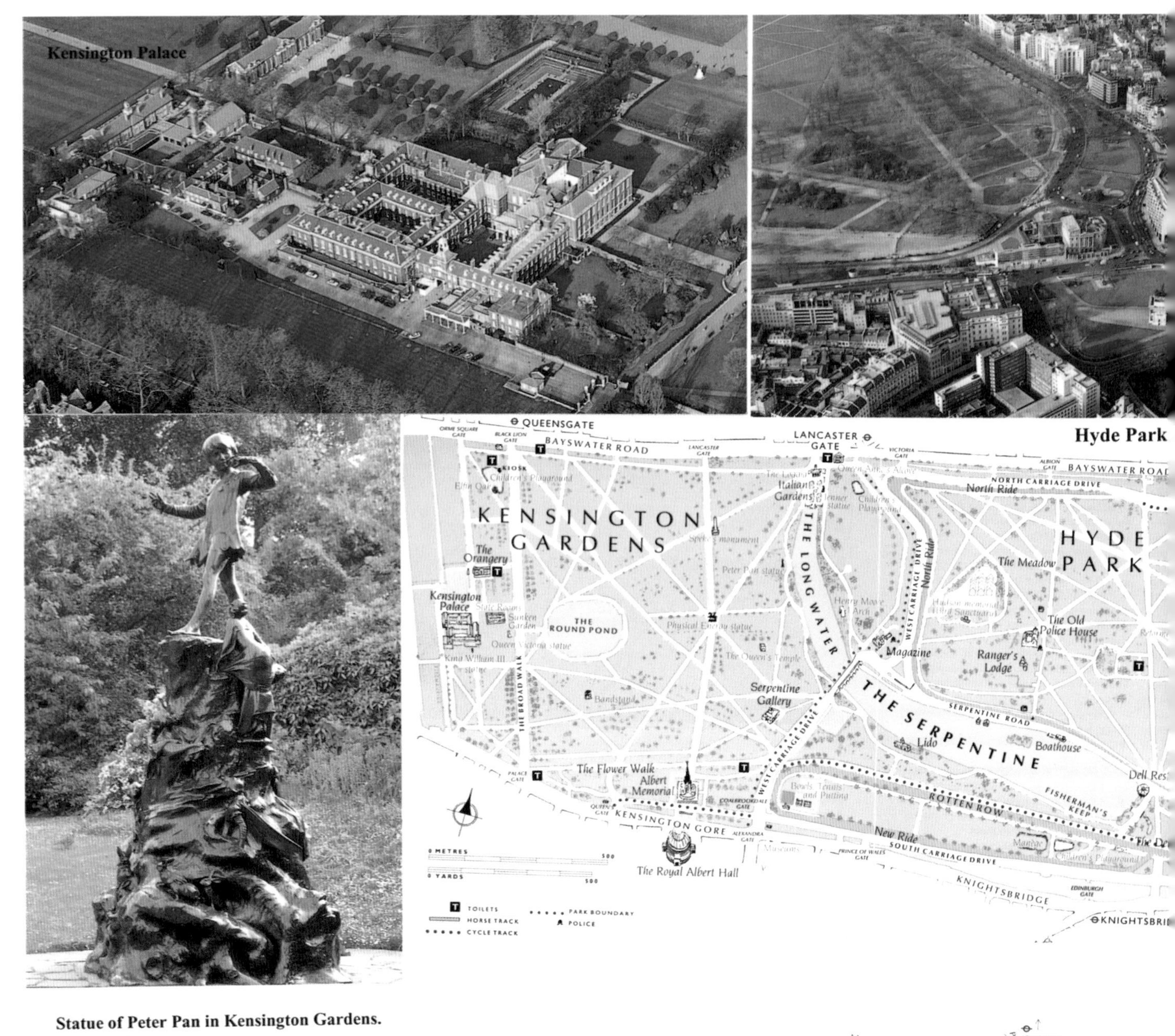

Statue of Peter Pan in Kensington Gardens.

Aerial view of London Zoo in Regents Park.

nd Hilton Hotel.

Horse riding in Hyde Park.

Aerial view showing Westminster, St James Park, Green Park and Hyde Park.

Maps Copyright Royal Parks

London Royal Parks and Zoo

These delightful parks are unique and not matched anywhere else in the world. St James Park is the oldest Royal Park, offering delight and relaxation to walkers. Henry VIII acquired the park as a hunting place for deer in 1532 and built the Palace of St James alongside. When his daughter, Elizabeth I, came to the throne she indulged her love of pageantry and pomp here. The lake supports a thriving number of wild birds. The adjacent Green Park with its mature trees contributes so much to Londoners' life. It is an important link between St James Park and Hyde Park forming a chain of open spaces.

Hyde Park is one of the capital's finest landscapes, extending over 140 hectares, and has been a Royal Park since 1536. Kensington Gardens with their magnificent trees are the setting for Kensington Palace, previously the home of William III and Mary II for their London home. Victoria spent much of her childhood there, and more recently the Princess of Wales, Diana, lived there until her tragic death in 1997. There is the peace of the Italian Gardens within a short distance of the palace. Children play with model boats on the Round Pond gathered normally at the Peter Pan statue.

The Regents Park is the home of the renowned Queen Mary's Rose Garden with an open-air theatre, café and London Zoo, which is one of the largest in the world. The Queen Mary gardens date back to the early 1930s. Within the northern boundary of the park a golf and tennis school has been established.

Other Royal Parks and gardens around London include Richmond Park, Hampton Court Gardens and Bushy Park.

Entrance to the Earth Galleries at the Natural History Museum, South Kensington, and a model of a dinosaur.

Portrait of King George III, 1761.

Silver Microscope made for George III (Science Museum, South Kensington).

Ancient Egyptian sculptures (British Museum, Bloomsbury).

Ludgate Hill looking towards St Paul's Cathedral, one of many paintings at the Museum of London, Barbican.

Early 1900 tram and conductor, Museum of London Transport in Covent Garden.

Queen Victoria's Military Uniform was shown at Victoria and Albert Museum, South Kensington.

London Museums

Great Britain has more museums to the square kilometre than any other country in the world, most of which are located in London. The greatest have paintings, ancient treasures, furniture, porcelain, metalwork and jewellery worth billions. Exceptional items include the Alfred Jewel, the Portland Vase and the Great Bed of Ware.

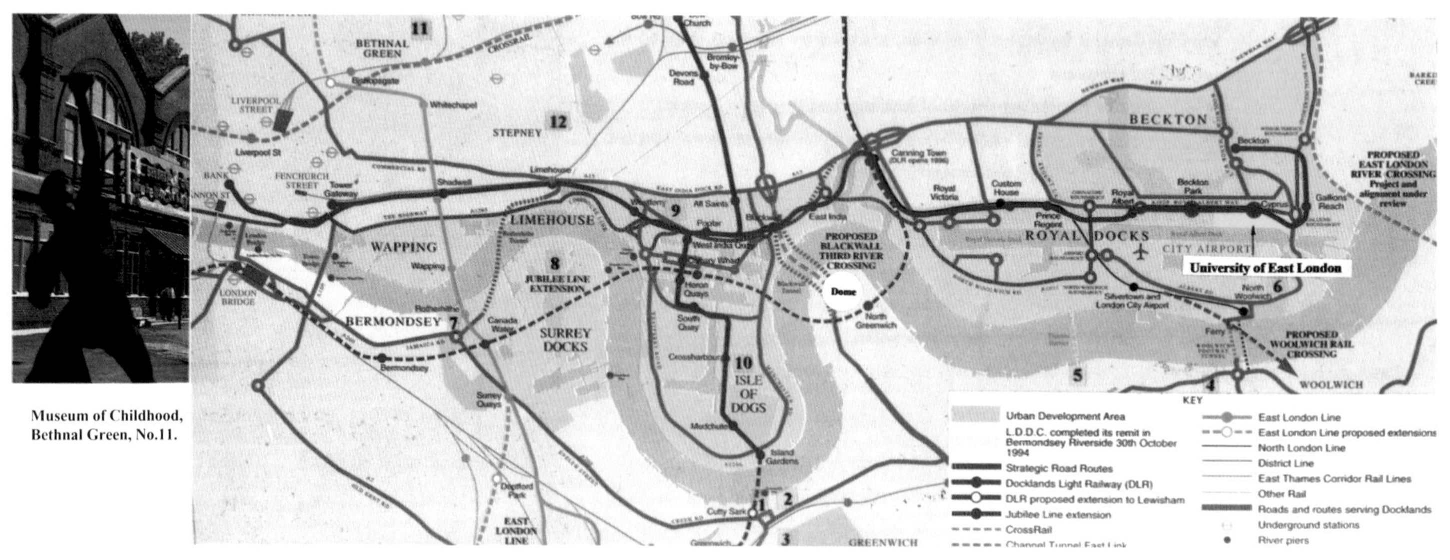

Museum of Childhood, Bethnal Green, No.11.

Museums in Greenwich and Docklands

Cutty Sark, No.1; National Maritime Museum, No.2; Old Royal Observatory, No.3; Royal Artillery Museum, No.4; Thames Barrier Visitor Centre, No.5; North Woolwich Railway Museum, No.6; Brunel's Engine House, No.7; Lavender Pond Pump House, No.8; Museum of Docklands, No.9; Island History Trust, No.10; Museum of Childhood, No.11; and Ragged School Museum, No.12.

The Great British Public House

Ask any Englishman what are the things he most likes and pretty high up on his list will come the pub. Where the word "bar" is internationally understood in America and other parts of the world, the notion of a public house is peculiarly Anglo-Saxon. For a street with inn signs is also an art gallery and a portrait of English life.

The Romans introduced the first signs. A bush hanging outside their tavernas represented the grapevine. The Saxon alehouses used an ale garland – a wreath hanging from a pole. Then came the painted board in 1393 when Richard II regularised the practice. Throughout history, the swinging pub signs have conveyed history and not just the beer inside that is an English institution.

It has been said that it was the church which created the inns. The first real inns were built on the drovers roads and pilgrims routes for people travelling to shrines. In medieval times, the masons and craftsmen imported to a town or village to work on building the church, had to live in these inns. That is why churches and pubs can stand close to each other and are often regarded as twin pillars of village life.

The British Public House is an oasis of civilisation and can be found in all parts of the country. Roaring log fires, oak beams, and well-tended gardens are everywhere. Many are dotted around London, full of informality and hospitality. A few of them have guestrooms. Pubs are great and refreshing places and offer a refuge from the pressures of modern living.

Above: The City Pride Public House, Isle of Dogs.
Below: Prospect of Whitby at Wapping.

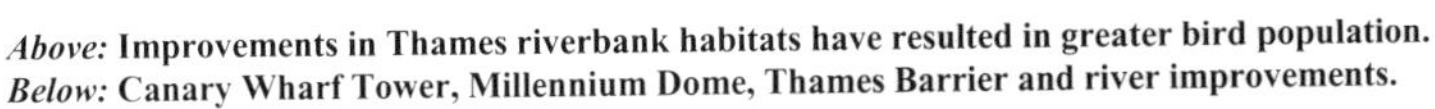

Above: Improvements in Thames riverbank habitats have resulted in greater bird population.
Below: Canary Wharf Tower, Millennium Dome, Thames Barrier and river improvements.

The Thames Barrier, the largest moveable flood barrier in the world built in 1984, prevents London from flooding.

River bus service for new Millennium, passing the Tower of London.

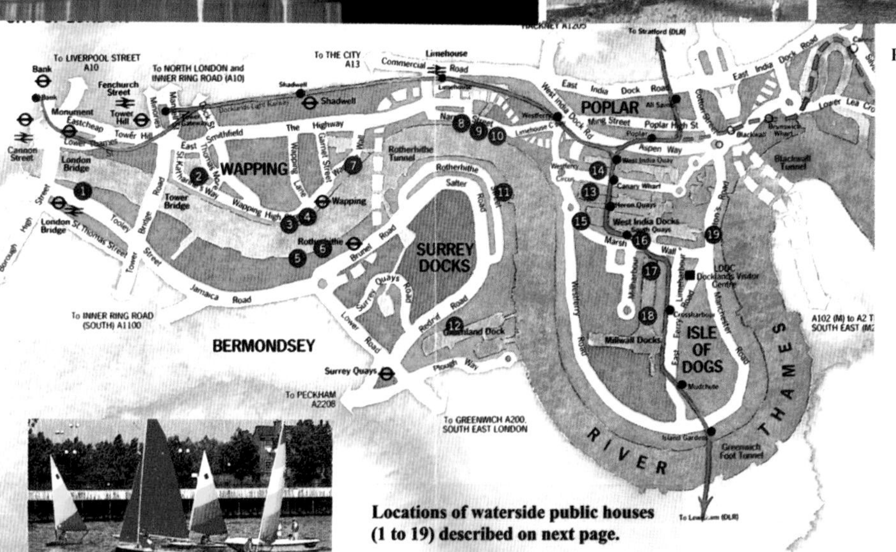

Locations of waterside public houses (1 to 19) described on next page.

River Thames

There is no finer or more suitable way of marking the Millennium in London than by making our river once again the stately concourse it was for centuries. Many groups have come together to make this happen. Spanned by bridges both new and historic, the Thames in the year 2000 is the thread running through a spectacular range of activities. Our river is again the pride of the City and a model for other nations.

Since the 1960s the River Thames has been given a new lease of life. Considerable effort in cleaning up the river has supported a remarkable recovery in the ecology of the river and its riverside habitats. Improvements in water quality created better waterside conditions where the riverbank ecology can be sustained. The reduction in commercial river traffic has made many wharf sides available for redevelopment and environmental enhancement schemes.

1

The Horniman at Hay's

A new pub built within Hay's Galleria, a magnificent riverside arcade with shopping and eating facilities. The Horniman is large and friendly with a busy lively atmosphere, especially in the evenings. Bar food is served at lunchtimes with tea and coffee also available.

Hay's Galleria, Tooley Street, SE1.
Telephone 071 407 3611
Opening times:
Mon-Fri 11.00am - 11.00pm
Sat 11.00am - 3.00pm
Sun 12.00 - 3.00pm

2

The Dickens Inn

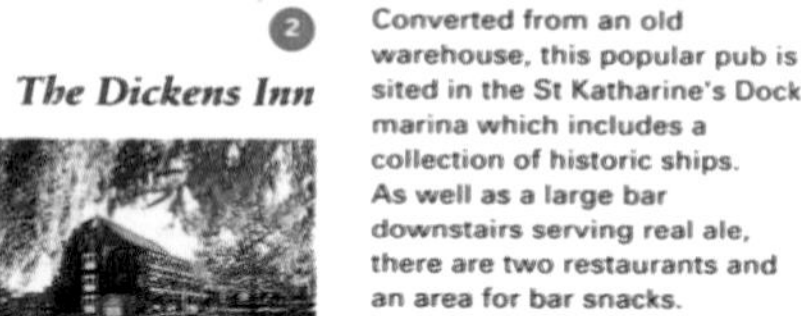

Converted from an old warehouse, this popular pub is sited in the St Katharine's Dock marina which includes a collection of historic ships. As well as a large bar downstairs serving real ale, there are two restaurants and an area for bar snacks.

St. Katharine's Way, E1
Telephone 071 488 1226
Opening times:
Restaurants: Mon-Sun 12.00am - 3.00pm, 6.30pm - 10.30pm. **Bar: Mon-Sat** 11.00am - 11.00pm. **Sun** 12.00am - 3.00pm, 7.00pm - 10.30pm

3

The Town of Ramsgate

Dwarfed by Olivers Wharf warehouses next to it, this pub lies in the heart of the Wapping Pierhead Conservation area. It has a long history of its own, as it is reputedly the spot where the infamous "Hanging" Judge Jeffries was caught whilst trying to flee the country. A cosy atmosphere prevails in this friendly pub, with bar meals available at lunchtimes and evenings.

62 Wapping High St, E1.
Telephone 071 488 2685
Opening times
Mon-Thur 12.00am - 3.00pm, 5.00pm - 11.00pm. **Fri** 12.00am - 11.00pm. **Sat-Sun** 12.00am - 3.00pm, 7.00pm - 11.00pm

4

The Captain Kidd

One of the most recently opened pubs on the river, this pleasant warehouse conversion has coveted window seats and an attractive riverside beer garden. The downstairs bar serves sandwiches, the first floor offers more substantial snacks and on the top floor there's a restaurant.

108 Wapping High Street, E1
Telephone 071 480 5759
Opening times
Restaurant: Tues-Sun 12.00am - 3.00pm, 6.30pm - 12.00pm. **Bar: Mon-Sat** 11.00am - 11.00pm. **Sun** 12.00am - 3.00pm, 7.00pm - 10.30pm

5

The Angel

An historic pub built near the site of King Edward III's manor house and said to be a favourite drinking spot of Captain Cook. The downstairs bar where snacks are served at lunchtimes is full of character with a balcony overlooking the river. Upstairs is a high quality restaurant serving at lunchtimes and evenings.

101 Bermondsey Wall East, Rotherhithe, SE16.
Telephone 071 237 3608
Opening times: Restaurant: Mon-Fri 12.00am - 3.00pm, 7.00pm - 10.00pm. **Sat** 7.00pm - 10.00pm. **Bar: Mon-Sat** 11.00am - 3.00pm, 5.30pm - 11.00pm. **Sun** 12.00am - 3.00pm, 7.00pm - 10.30pm.

6

The Mayflower

Named after the ship that took the Pilgrim Fathers to America, this is another historic pub with lots of atmosphere in its small downstairs bars. Bar food is served throughout the pub with a seating area upstairs affording excellent river views.

117 Rotherhithe Street, SE16
Telephone 071 237 4088
Opening Times:
Mon-Fri 12.00am - 3.00pm, 6.00pm - 11.00pm. **Sat** 12.00am - 3.00pm, 6.30pm - 11.00pm. **Sun** 12.00am -

7

The Prospect of Whitby

Probably the most famous pub on the river, dating back to the sixteenth century and boasting Samuel Pepys and Turner the painter amongst its former regulars. The downstairs bar with its original flagstone floor has a good variety of bar meals and a beer garden overlooking the river. Upstairs the restaurant is open for lunches and dinner.

5 Wapping Wall, E1
Telephone 071 481 1095
Opening times:
Restaurant : Mon-Fri 12.00am - 2.00pm, 7.00pm - 10.00pm. **Sat** 7.00pm - 10.00pm. **Sun** 12.00am - 2.00pm. **Bar: Mon-Thur** 11.30am - 3.00pm,

8

The Barleymow

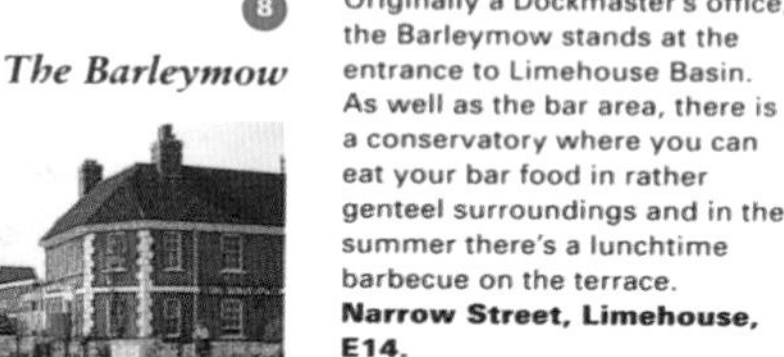

Originally a Dockmaster's office, the Barleymow stands at the entrance to Limehouse Basin. As well as the bar area, there is a conservatory where you can eat your bar food in rather genteel surroundings and in the summer there's a lunchtime barbecue on the terrace.

Narrow Street, Limehouse, E14.
Telephone 071 265 8931
Opening times:
Restaurant: Mon 12.00am - 2.00pm **Tue-Fri** 12.00am - 2.00pm, 7.00pm - 9.00pm. **Sat** 7.00pm - 9.00pm **Bar: Mon-Fri** 11.30am - 3.00pm, 5.30pm - 11.00pm. **Sat** 12.00am - 3.00pm, 7.00pm - 11.00pm. **Sun** 12.00am - 3.00pm, 7.00pm - 10.30pm

9

The Grapes

Famous for its Dickens connection as the model for the inn from Our Mutual Friend, the Grapes certainly feels old and atmospheric with its narrow bar and old river balcony. Get here early at lunchtimes to secure a seat. Like many of the other pubs along the river, a restaurant can be found upstairs, this one specialising in seafood.

76 Narrow Street, E14
Telephone 071 987 4396
Opening times
Restaurant - Mon-Fri 12.00am - 2.00pm. **Tues-Sat** 7.00pm - 9.00pm
Bar: Mon-Sun 12.00am - 2.30pm
Mon-Fri 5.30pm - 11.00pm.
Sat 7.00pm - 11.00pm.
Sun 7.00pm - 10.30pm

10

Bootys

Right next to its famous neighbour, the Grapes,in the heart of the Narrow Street conservation area, this popular bar offers home cooked food all day up until 9.30pm. A cosy atmosphere and friendly staff complement the Olde Worlde feeling inside.

92A Narrow Street, E14
Telephone 071 987 8343
Opening times
Mon-Fri 11.00am - 11.00pm
Sat 11.00am - 3.00pm, 7.00pm - 12.00pm. **Sun** 12.00am - 2.30pm

11

Scandic Crown Hotel

The first new hotel in London Docklands offers all day drinking at the Copenhagen bar, where French windows open out onto the riverside. There is an impressive view of Canary Wharf across the Thames. Tea and coffee are also available.

Scandic Crown Hotel Nelson Dock, Rotherhithe Street, SE16
Telephone 071 231 1001
Opening times
Mon-Fri 10.00am - 12.00pm

12

The Moby Dick

A new pub which overlooks Greenland Dock, the largest of the Surrey Docks and home to many award-winning developments. Relax and watch members of the local watersports club going through their paces on the dock. Bar food is served at lunchtimes and evenings.

6 Russell Place, South Dock, SE16
Telephone 071 231 5482
Opening times:
Mon-Wed 12.00am - 3.00pm, 5.00pm - 11.00pm. **Thur** 12.00am - 11.00pm. **Fri-Sat** 12.00am - 12.00pm. **Sun** 12.00am - 3.00pm, 7.00pm - 10.30pm

13

The Henry Addington

One of the newest venues in London Docklands is the Henry Addington at Canary Wharf, an American style bar which already conveys a happy well-established atmosphere. An unremarkable entrance at the corner of the American Express building leads to the large bar and seating area. Choose from sandwiches or upmarket bar snacks at lunchtimes, with a more substantial menu in the evenings.

20-28 Mackenzie Walk, Canary Wharf, E14
Telephone 071 512 9022
Opening times
Mon-Fri 11.00am - 11.00pm
Sat-Sun 11.30am - 4.00pm

14

The Cat & the Canary

Opening in May 1992, the latest pub in Docklands looks over the West India Dock towards the magnificent Port East complex of sugar warehouses. Sit out-side in fine weather or enjoy the olde worlde atmosphere indoors with Real Ale. Bar food served.

Fisherman's Walk
Canary Wharf
London E14
Opening times
Mon - Fri 11.00 - 9.00pm
Sat 11.00 - 5.00pm
Sun 12.00 - 3.00pm

15

Drummonds

A large and stylish bar on the West India Dock at Heron Quays, part of a national chain of cafe bars. Enjoy the outside terrace on the dock edge in good weather. Food is served during the day.

Heron Quays, Marsh Wall, E14
Telephone 071 538 3357
Opening times
Mon-Fri 8.00am - 9.00pm
Not open at weekends

16

The Waterfront

A modern and lively bar overlooking West India Dock and with a great view of Canary Wharf, this pub is on two levels. The lower offers bar food at lunchtimes or just a pleasant place for a drink. Upstairs you'll find a high quality restaurant, lunchtimes only.

South Quay Plaza, 187 Marsh Wall, E14. Telephone 071 537 2823
Opening times
Restaurant : Mon-Fri 12.00am - 3.00pm

17

The Spinnaker

A light, airy bar, part of a new development, Harbour Island, which stands in Millwall Dock. A popular place for a lunchtime drink with workers from the surrounding offices. Tea and coffee are served all day and bar food is available at lunchtimes.

Harbour Island, Harbour Exchange Square, E14
Telephone 071 538 9329
Opening times
Mon 11.30am - 8.00pm
Tue-Fri 11.00am - 8.00pm

18

Manzis

This branch of the famous Leicester Square fish restaurant also has a bar upstairs with excellent views over Millwall Dock, and a balcony to enjoy sitting out in warm weather. Italian food is served at the bar all day.

Turnberry Quay, London E14
Telephone 071 538 9615
Opening times
Restaurant: Mon - Fri 12.00 am - 3.00pm, 6.00pm - 11pm.

19

The Gun

A bit off the beaten track, this old pub is nevertheless worth a visit. Steeped in the history of the Coldharbour conservation area, this is said to be the tavern used by Nelson for his rendezvous with Lady Hamilton. Enjoy lunchtime barfood on the terrace overhanging the Thames.

Coldharbour, E14
Telephone 071 987 1692
Opening times
Mon-Sat 11.00am - 3.00pm, 5.00pm - 11.00pm. **Sun** 12.00am - 3.00pm,

For locations see map on previous page.

River Thames Waterside Public Houses

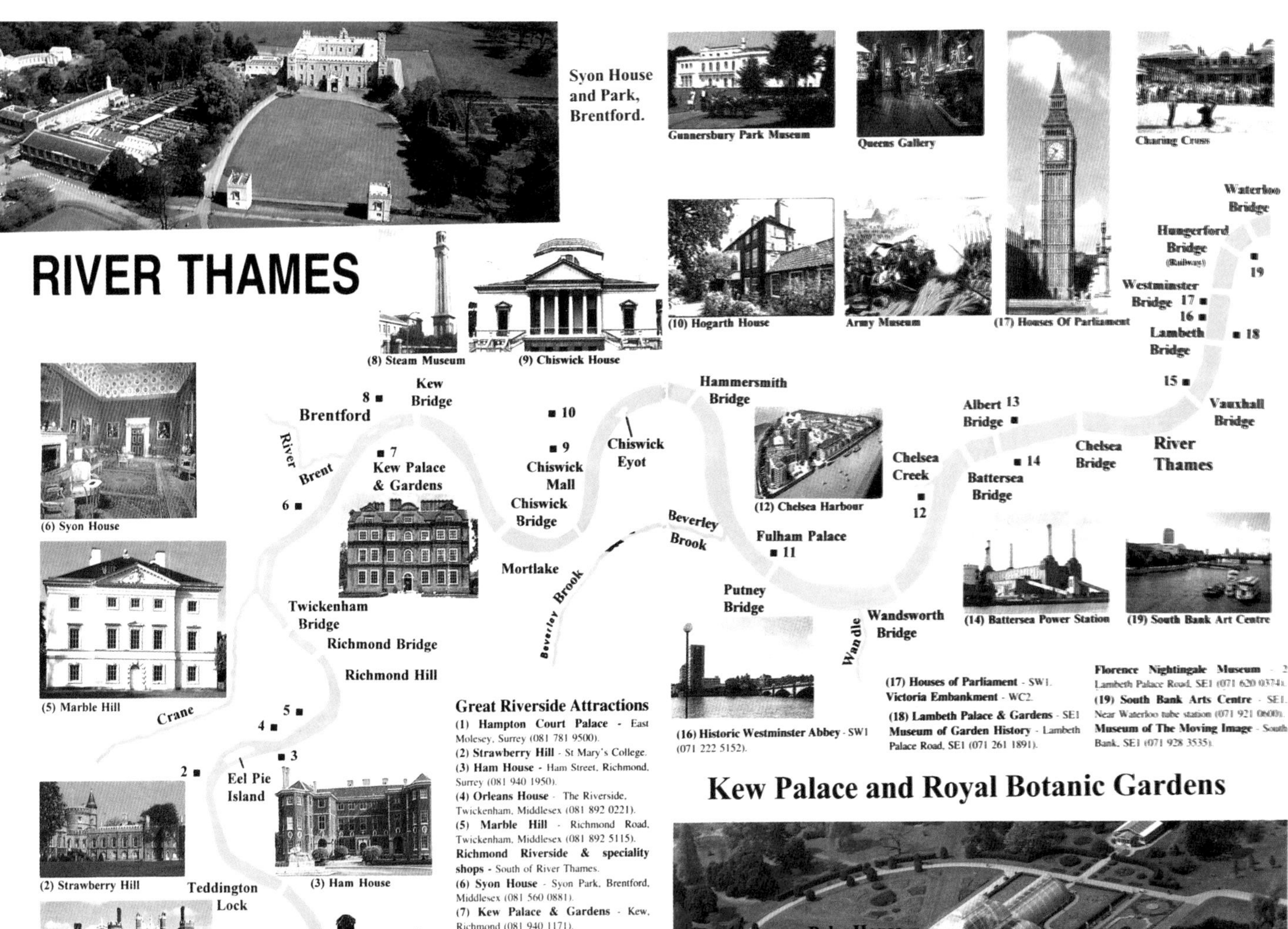

Great Riverside Attractions

(1) Hampton Court Palace - East Molesey, Surrey (081 781 9500).
(2) Strawberry Hill - St Mary's College.
(3) Ham House - Ham Street, Richmond, Surrey (081 940 1950).
(4) Orleans House - The Riverside, Twickenham, Middlesex (081 892 0221).
(5) Marble Hill - Richmond Road, Twickenham, Middlesex (081 892 5115).
Richmond Riverside & speciality shops - South of River Thames.
(6) Syon House - Syon Park, Brentford, Middlesex (081 560 0881).
(7) Kew Palace & Gardens - Kew, Richmond (081 940 1171).
(8) Kew Bridge Steam Museum - Green Dragon Lane, Brentford (081 568 4757).
(9) Chiswick House - Burlington Lane W4 (081 995 0508).
(10) Hogarth House - Hogarth Lane, W4 (081 994 6757).
Gunnersbury Park Museum - Acton W3 (081 993 1612).
(11) Fulham Palace - Bishops Avenue, Fulham, SW6 (071 736 5821).
(12) Chelsea Harbour - London SW10
(13) Chelsea Royal Hospital - Chelsea, SW3 (071 730 0161).
(14) Battersea Park & Power Station - SW19 (081 871 7530).
(15) Tate Gallery - The Millbank, SW1 (071 887 8000).

Cruise from Westminster to Hampton Court

Kew Palace and Royal Botanic Gardens

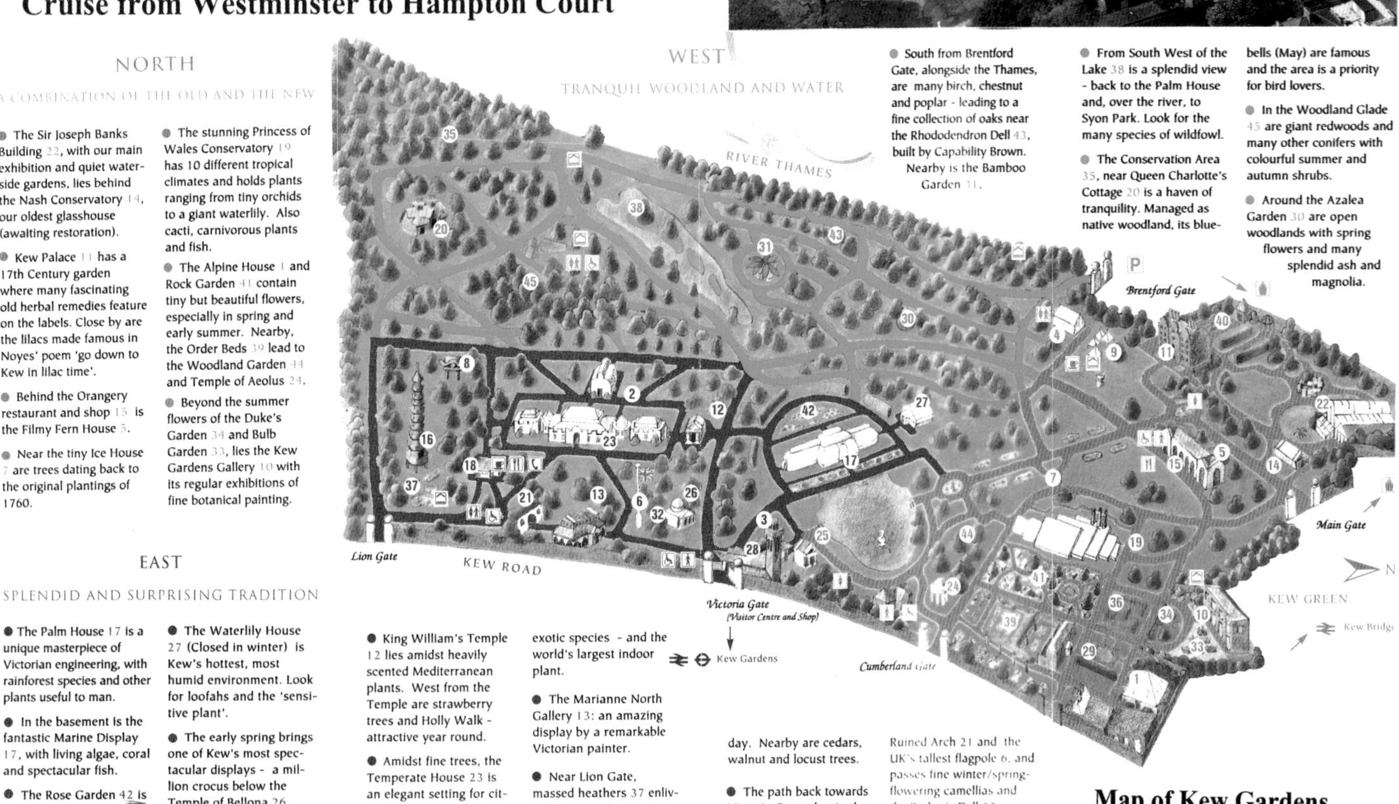

NORTH

A COMBINATION OF THE OLD AND THE NEW

- The Sir Joseph Banks Building 22, with our main exhibition and quiet water-side gardens, lies behind the Nash Conservatory 14, our oldest glasshouse (awaiting restoration).
- Kew Palace 11 has a 17th Century garden where many fascinating old herbal remedies feature on the labels. Close by are the lilacs made famous in Noyes' poem 'go down to Kew in lilac time'.
- Behind the Orangery restaurant and shop 15 is the Filmy Fern House 5.
- Near the tiny Ice House 7 are trees dating back to the original plantings of 1760.
- The stunning Princess of Wales Conservatory 19 has 10 different tropical climates and holds plants ranging from tiny orchids to a giant waterlily. Also cacti, carnivorous plants and fish.
- The Alpine House 1 and Rock Garden 41 contain tiny but beautiful flowers, especially in spring and early summer. Nearby, the Order Beds 39 lead to the Woodland Garden 44 and Temple of Aeolus 24.
- Beyond the summer flowers of the Duke's Garden 34 and Bulb Garden 33, lies the Kew Gardens Gallery 10 with its regular exhibitions of fine botanical painting.

WEST

TRANQUIL WOODLAND AND WATER

- South from Brentford Gate, alongside the Thames, are many birch, chestnut and poplar - leading to a fine collection of oaks near the Rhododendron Dell 43, built by Capability Brown. Nearby is the Bamboo Garden 31.
- From South West of the Lake 38 is a splendid view - back to the Palm House and, over the river, to Syon Park. Look for the many species of wildfowl.
- The Conservation Area 35, near Queen Charlotte's Cottage 20 is a haven of tranquility. Managed as native woodland, its bluebells (May) are famous and the area is a priority for bird lovers.
- In the Woodland Glade 45 are giant redwoods and many other conifers with colourful summer and autumn shrubs.
- Around the Azalea Garden 30 are open woodlands with spring flowers and many splendid ash and magnolia.

EAST

SPLENDID AND SURPRISING TRADITION

- The Palm House 17 is a unique masterpiece of Victorian engineering, with rainforest species and other plants useful to man.
- In the basement is the fantastic Marine Display 17, with living algae, coral and spectacular fish.
- The Rose Garden 42 is a summer feature.
- The Waterlily House 27 (Closed in winter) is Kew's hottest, most humid environment. Look for loofahs and the 'sensitive plant'.
- The early spring brings one of Kew's most spectacular displays - a million crocus below the Temple of Bellona 26.
- King William's Temple 12 lies amidst heavily scented Mediterranean plants. West from the Temple are strawberry trees and Holly Walk - attractive year round.
- Amidst fine trees, the Temperate House 23 is an elegant setting for citrus fruits, tea and many exotic species - and the world's largest indoor plant.
- The Marianne North Gallery 13: an amazing display by a remarkable Victorian painter.
- Near Lion Gate, massed heathers 37 enliven the greyest winter's day. Nearby are cedars, walnut and locust trees.
- The path back towards Victoria Gate takes in the Ruined Arch 21 and the UK's tallest flagpole 6, and passes fine winter/spring-flowering camellias and the Berberis Dell 32.

Map of Kew Gardens

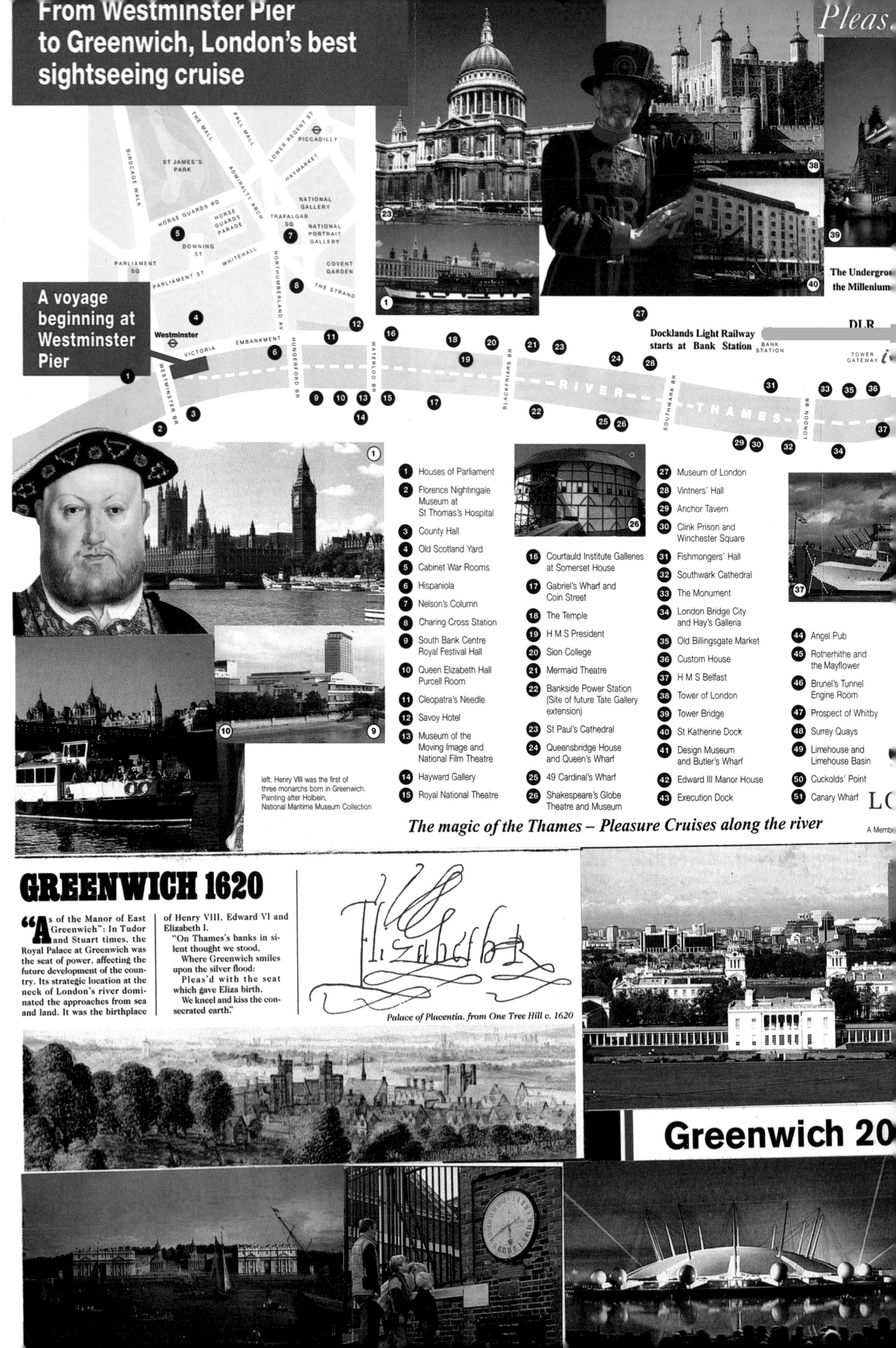

From Westminster Pier to Greenwich, London's best sightseeing cruise

A voyage beginning at Westminster Pier

1. Houses of Parliament
2. Florence Nightingale Museum at St Thomas's Hospital
3. County Hall
4. Old Scotland Yard
5. Cabinet War Rooms
6. Hispaniola
7. Nelson's Column
8. Charing Cross Station
9. South Bank Centre Royal Festival Hall
10. Queen Elizabeth Hall Purcell Room
11. Cleopatra's Needle
12. Savoy Hotel
13. Museum of the Moving Image and National Film Theatre
14. Hayward Gallery
15. Royal National Theatre
16. Courtauld Institute Galleries at Somerset House
17. Gabriel's Wharf and Coin Street
18. The Temple
19. H M S President
20. Sion College
21. Mermaid Theatre
22. Bankside Power Station (Site of future Tate Gallery extension)
23. St Paul's Cathedral
24. Queensbridge House and Queen's Wharf
25. 49 Cardinal's Wharf
26. Shakespeare's Globe Theatre and Museum
27. Museum of London
28. Vintners' Hall
29. Anchor Tavern
30. Clink Prison and Winchester Square
31. Fishmongers' Hall
32. Southwark Cathedral
33. The Monument
34. London Bridge City and Hay's Galleria
35. Old Billingsgate Market
36. Custom House
37. H M S Belfast
38. Tower of London
39. Tower Bridge
40. St Katherine Dock
41. Design Museum and Butler's Wharf
42. Edward III Manor House
43. Execution Dock
44. Angel Pub
45. Rotherhithe and the Mayflower
46. Brunel's Tunnel Engine Room
47. Prospect of Whitby
48. Surrey Quays
49. Limehouse and Limehouse Basin
50. Cuckolds' Point
51. Canary Wharf

left: Henry VIII was the first of three monarchs born in Greenwich. Painting after Holbein, National Maritime Museum Collection

The magic of the Thames – Pleasure Cruises along the river

GREENWICH 1620

"As of the Manor of East Greenwich": In Tudor and Stuart times, the Royal Palace at Greenwich was the seat of power, affecting the future development of the country. Its strategic location at the neck of London's river dominated the approaches from sea and land. It was the birthplace of Henry VIII, Edward VI and Elizabeth I.

"On Thames's banks in silent thought we stood,
Where Greenwich smiles upon the silver flood:
Pleas'd with the seat which gave Eliza birth,
We kneel and kiss the consecrated earth."

Palace of Placentia, from One Tree Hill c. 1620

Greenwich 20

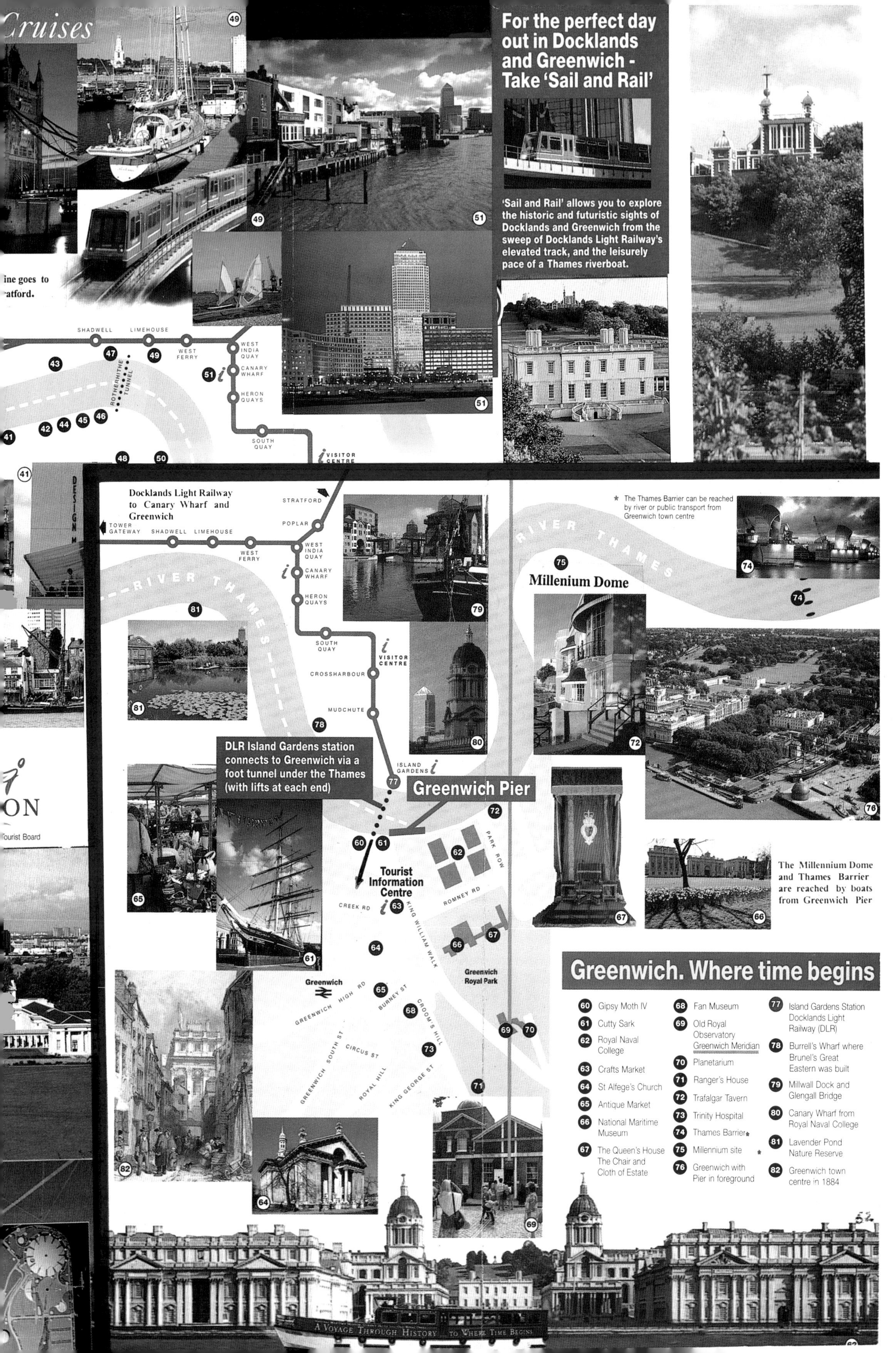
Cruises
ine goes to
atford.
For the perfect day out in Docklands and Greenwich - Take 'Sail and Rail'
'Sail and Rail' allows you to explore the historic and futuristic sights of Docklands and Greenwich from the sweep of Docklands Light Railway's elevated track, and the leisurely pace of a Thames riverboat.
SHADWELL
LIMEHOUSE
WEST FERRY
WEST INDIA QUAY
CANARY WHARF
HERON QUAYS
SOUTH QUAY
ROTHERHITHE TUNNEL
VISITOR CENTRE
ON
Tourist Board
Docklands Light Railway to Canary Wharf and Greenwich
STRATFORD
POPLAR
TOWER GATEWAY
CROSSHARBOUR
MUDCHUTE
ISLAND GARDENS
RIVER THAMES
* The Thames Barrier can be reached by river or public transport from Greenwich town centre
Millenium Dome
DLR Island Gardens station connects to Greenwich via a foot tunnel under the Thames (with lifts at each end)
Greenwich Pier
Tourist Information Centre
CREEK RD
KING WILLIAM WALK
ROMNEY RD
PARK ROW
Greenwich
GREENWICH HIGH RD
BURNEY ST
CROOM'S HILL
CIRCUS ST
GREENWICH SOUTH ST
ROYAL HILL
KING GEORGE ST
Greenwich Royal Park
The Millennium Dome and Thames Barrier are reached by boats from Greenwich Pier
Greenwich. Where time begins
60 Gipsy Moth IV
61 Cutty Sark
62 Royal Naval College
63 Crafts Market
64 St Alfege's Church
65 Antique Market
66 National Maritime Museum
67 The Queen's House The Chair and Cloth of Estate
68 Fan Museum
69 Old Royal Observatory Greenwich Meridian
70 Planetarium
71 Ranger's House
72 Trafalgar Tavern
73 Trinity Hospital
74 Thames Barrier*
75 Millennium site
76 Greenwich with Pier in foreground
77 Island Gardens Station Docklands Light Railway (DLR)
78 Burrell's Wharf where Brunel's Great Eastern was built
79 Millwall Dock and Glengall Bridge
80 Canary Wharf from Royal Naval College
81 Lavender Pond Nature Reserve
82 Greenwich town centre in 1884
A VOYAGE THROUGH HISTORY TO WHERE TIME BEGINS

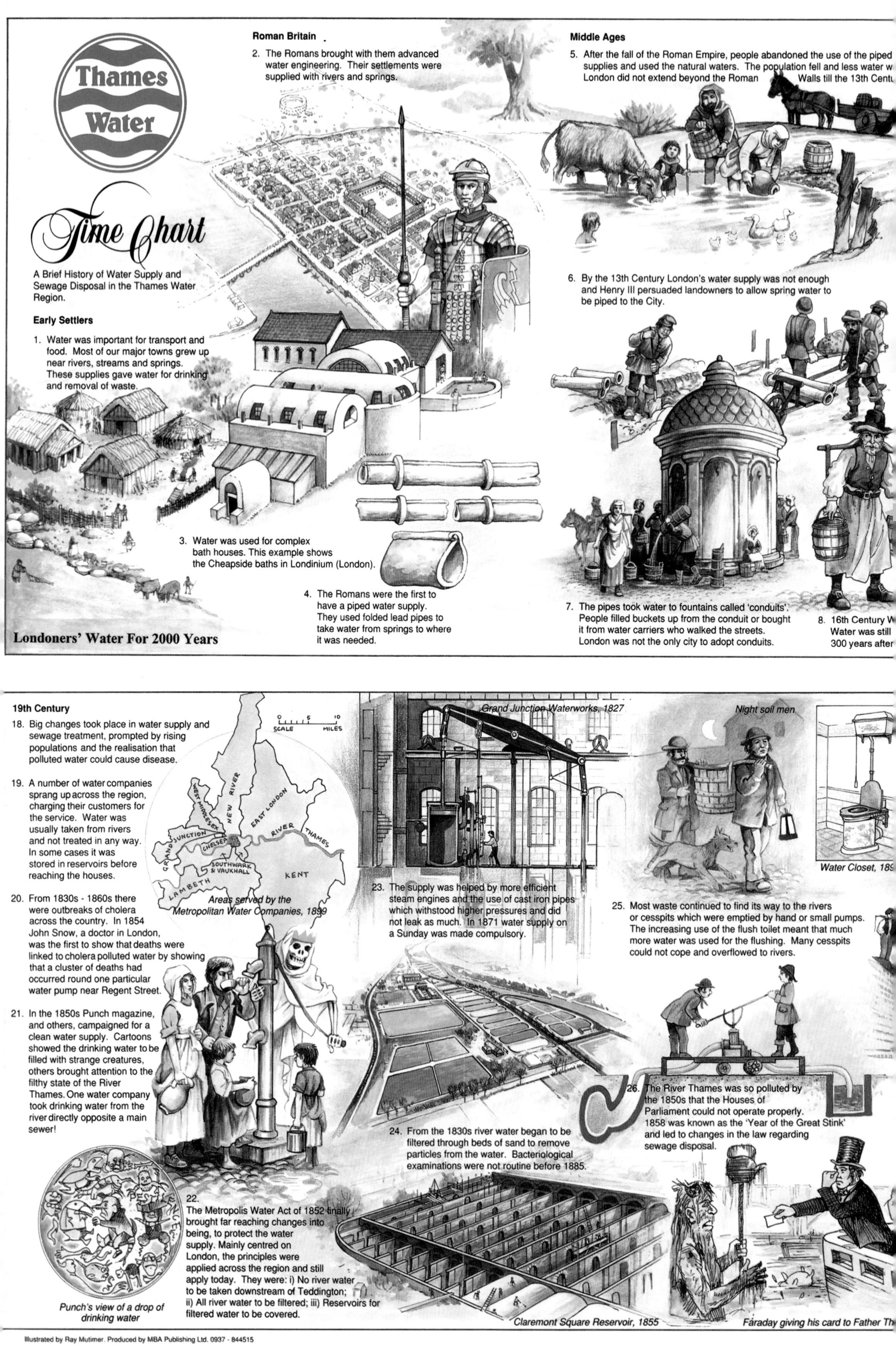

Thames Water
Time Chart
A Brief History of Water Supply and Sewage Disposal in the Thames Water Region.
Early Settlers
1. Water was important for transport and food. Most of our major towns grew up near rivers, streams and springs. These supplies gave water for drinking and removal of waste.
Roman Britain
2. The Romans brought with them advanced water engineering. Their settlements were supplied with rivers and springs.
3. Water was used for complex bath houses. This example shows the Cheapside baths in Londinium (London).
4. The Romans were the first to have a piped water supply. They used folded lead pipes to take water from springs to where it was needed.
Middle Ages
5. After the fall of the Roman Empire, people abandoned the use of the piped supplies and used the natural waters. The population fell and less water w London did not extend beyond the Roman Walls till the 13th Centu
6. By the 13th Century London's water supply was not enough and Henry III persuaded landowners to allow spring water to be piped to the City.
7. The pipes took water to fountains called 'conduits'. People filled buckets up from the conduit or bought it from water carriers who walked the streets. London was not the only city to adopt conduits.
8. 16th Century W Water was still 300 years after
Londoners' Water For 2000 Years
19th Century
18. Big changes took place in water supply and sewage treatment, prompted by rising populations and the realisation that polluted water could cause disease.
19. A number of water companies sprang up across the region, charging their customers for the service. Water was usually taken from rivers and not treated in any way. In some cases it was stored in reservoirs before reaching the houses.
20. From 1830s - 1860s there were outbreaks of cholera across the country. In 1854 John Snow, a doctor in London, was the first to show that deaths were linked to cholera polluted water by showing that a cluster of deaths had occurred round one particular water pump near Regent Street.
21. In the 1850s Punch magazine, and others, campaigned for a clean water supply. Cartoons showed the drinking water to be filled with strange creatures, others brought attention to the filthy state of the River Thames. One water company took drinking water from the river directly opposite a main sewer!
SCALE 0 5 10 MILES
WEST MIDDLESEX
NEW RIVER
EAST LONDON
GRAND JUNCTION
CHELSEA
RIVER THAMES
SOUTHWARK & VAUXHALL
LAMBETH
KENT
Areas served by the Metropolitan Water Companies, 1899
22. The Metropolis Water Act of 1852 finally brought far reaching changes into being, to protect the water supply. Mainly centred on London, the principles were applied across the region and still apply today. They were: i) No river water to be taken downstream of Teddington; ii) All river water to be filtered; iii) Reservoirs for filtered water to be covered.
Punch's view of a drop of drinking water
Grand Junction Waterworks, 1827
23. The supply was helped by more efficient steam engines and the use of cast iron pipes which withstood higher pressures and did not leak as much. In 1871 water supply on a Sunday was made compulsory.
24. From the 1830s river water began to be filtered through beds of sand to remove particles from the water. Bacteriological examinations were not routine before 1885.
Claremont Square Reservoir, 1855
Night soil men
Water Closet, 18
25. Most waste continued to find its way to the rivers or cesspits which were emptied by hand or small pumps. The increasing use of the flush toilet meant that much more water was used for the flushing. Many cesspits could not cope and overflowed to rivers.
26. The River Thames was so polluted by the 1850s that the Houses of Parliament could not operate properly. 1858 was known as the 'Year of the Great Stink' and led to changes in the law regarding sewage disposal.
Faraday giving his card to Father Th
Illustrated by Ray Mutimer. Produced by MBA Publishing Ltd. 0937 - 844515

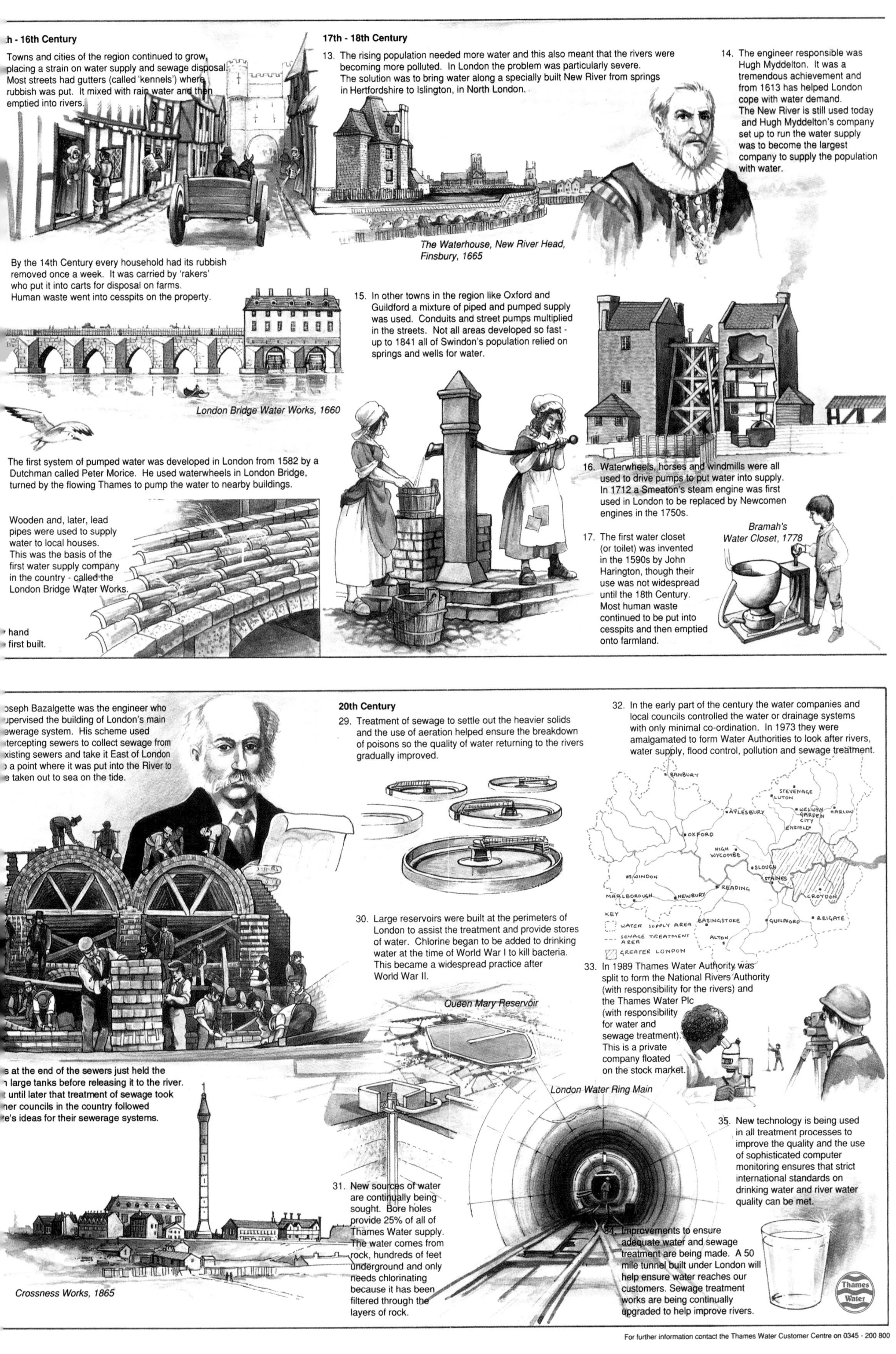

h - 16th Century
Towns and cities of the region continued to grow, placing a strain on water supply and sewage disposal. Most streets had gutters (called 'kennels') where rubbish was put. It mixed with rain water and then emptied into rivers.
By the 14th Century every household had its rubbish removed once a week. It was carried by 'rakers' who put it into carts for disposal on farms. Human waste went into cesspits on the property.
London Bridge Water Works, 1660
The first system of pumped water was developed in London from 1582 by a Dutchman called Peter Morice. He used waterwheels in London Bridge, turned by the flowing Thames to pump the water to nearby buildings.
Wooden and, later, lead pipes were used to supply water to local houses. This was the basis of the first water supply company in the country - called the London Bridge Water Works.
hand
first built.
17th - 18th Century
13. The rising population needed more water and this also meant that the rivers were becoming more polluted. In London the problem was particularly severe. The solution was to bring water along a specially built New River from springs in Hertfordshire to Islington, in North London.
The Waterhouse, New River Head, Finsbury, 1665
14. The engineer responsible was Hugh Myddelton. It was a tremendous achievement and from 1613 has helped London cope with water demand. The New River is still used today and Hugh Myddelton's company set up to run the water supply was to become the largest company to supply the population with water.
15. In other towns in the region like Oxford and Guildford a mixture of piped and pumped supply was used. Conduits and street pumps multiplied in the streets. Not all areas developed so fast - up to 1841 all of Swindon's population relied on springs and wells for water.
16. Waterwheels, horses and windmills were all used to drive pumps to put water into supply. In 1712 a Smeaton's steam engine was first used in London to be replaced by Newcomen engines in the 1750s.
17. The first water closet (or toilet) was invented in the 1590s by John Harington, though their use was not widespread until the 18th Century. Most human waste continued to be put into cesspits and then emptied onto farmland.
Bramah's Water Closet, 1778
oseph Bazalgette was the engineer who upervised the building of London's main ewerage system. His scheme used tercepting sewers to collect sewage from xisting sewers and take it East of London a point where it was put into the River to e taken out to sea on the tide.
s at the end of the sewers just held the n large tanks before releasing it to the river. until later that treatment of sewage took her councils in the country followed e's ideas for their sewerage systems.
Crossness Works, 1865
20th Century
29. Treatment of sewage to settle out the heavier solids and the use of aeration helped ensure the breakdown of poisons so the quality of water returning to the rivers gradually improved.
30. Large reservoirs were built at the perimeters of London to assist the treatment and provide stores of water. Chlorine began to be added to drinking water at the time of World War I to kill bacteria. This became a widespread practice after World War II.
Queen Mary Reservoir
31. New sources of water are continually being sought. Bore holes provide 25% of all of Thames Water supply. The water comes from rock, hundreds of feet underground and only needs chlorinating because it has been filtered through the layers of rock.
32. In the early part of the century the water companies and local councils controlled the water or drainage systems with only minimal co-ordination. In 1973 they were amalgamated to form Water Authorities to look after rivers, water supply, flood control, pollution and sewage treatment.
BANBURY
STEVENAGE
LUTON
AYLESBURY
WELWYN GARDEN CITY
HARLOW
ENFIELD
OXFORD
HIGH WYCOMBE
SLOUGH
SWINDON
STAINES
READING
MARLBOROUGH
NEWBURY
CROYDON
BASINGSTOKE
GUILDFORD
REIGATE
ALTON
KEY
WATER SUPPLY AREA
SEWAGE TREATMENT AREA
GREATER LONDON
33. In 1989 Thames Water Authority was split to form the National Rivers Authority (with responsibility for the rivers) and the Thames Water Plc (with responsibility for water and sewage treatment). This is a private company floated on the stock market.
London Water Ring Main
34. Improvements to ensure adequate water and sewage treatment are being made. A 50 mile tunnel built under London will help ensure water reaches our customers. Sewage treatment works are being continually upgraded to help improve rivers.
35. New technology is being used in all treatment processes to improve the quality and the use of sophisticated computer monitoring ensures that strict international standards on drinking water and river water quality can be met.
Thames Water
For further information contact the Thames Water Customer Centre on 0345 - 200 800

IN AND AROUND LONDON

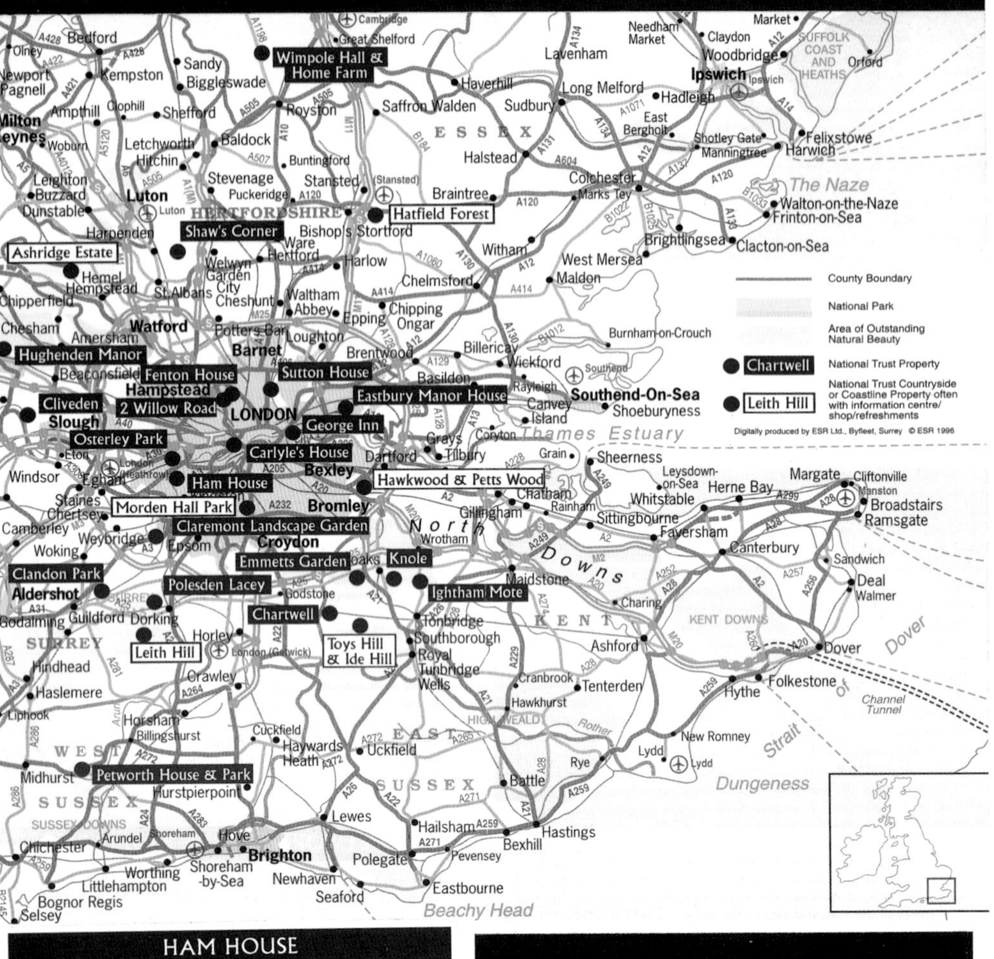

HAM HOUSE

Ham, Richmond TW10 7RS
Tel: 0181 940 1950

LOCATION: On S bank of Thames, W of A307, at Petersham *Station:* Richmond. *Underground:* 1½ml via Thames towpath, 2ml by road; Kingston 2ml. Also foot ferry from Marble Hill House.

Gloriously situated on the banks of the River Thames, Ham House is one of the finest 17th-century houses in Europe. Come and marvel the furniture, paintings and textiles of the extravagant Duchess of Lauderdale, much of her 1670s redecoration can still be seen.

THE GREEN ROOM AT HAM HOUSE

OPEN: House: 29 March to 2 Nov: Sat & Sun 12–5, Mon to Wed 1–5; last admission 4.30. **Garden:** open daily except Fri 10.30–6 (or dusk if earlier). Closed 25/26 Dec & 1 Jan.

ADMISSION: House: £4.50, family ticket £12. Groups welcome. Please book in advance.

HUGHENDEN MANOR

High Wycombe HP14 4LA
Tel: 01494 532580

LOCATION: 1½ml N of High Wycombe; on W side of the Great Missenden road (A4128). *Station:* High Wycombe 2ml.

Home of Queen Victoria's favourite, Prime Minister Benjamin Disraeli, the red brick Manor house with its 'gothic' skyline is set in colourful Victorian style gardens and surrounded by parkland and woodland. Inside, the house is filled with fascinating mementoes of Disraeli's life as statesman and writer, and with original furniture, pictures and books. The newly refurbished Victorian Stableblock holds the Coach House Shop, Stableyard Tea Room and an exhibition area.

OPEN: House: 1 to 30 March: Sat & Sun only; 31 March to end Oct: Wed to Sun & BH Mon 1–5. Last admission 4.30. Closed Good Fri. On BH weekends and other busy days entry is by timed ticket only. **Garden:** same days as house 12–5. **Park & woodland:** open all year.

ADMISSION: House & garden: £3.80; family ticket £9.50. Garden only £1, children 50p. Park & woodland free. Groups welcome. Please book in advance

HUGHENDEN MANOR

There are many other National Trust properties in the South East of England which are not featured in this guide.

IGHTHAM MOTE

Ivy Hatch, Sevenoaks TN15 0NT
Tel: 01732 810378

LOCATION: 6ml E of Sevenoaks, off A25, and 2½ml S of Ightham, off A227. *Station:* Borough Green & Wrotham 3½ml; Hildenborough 4ml

IGHTHAM MOTE

Explore 650 years of history in one house! Cross the walled moat and enter a medieval and Tudor manor house with rooms dating from as early as the 1340s. See the courtyard, the Great Hall, and the Robinson Library laid out as seen in a 1960 edition of Homes and Gardens magazine. Learn in a major exhibition about the restoration project - the largest of its kind ever undertaken by the National Trust - which is continuing in the chapel, the billiards room and the drawing room. See the methods and materials that are being used. Free introductory talks. Enjoy the extensive grounds - gardens, lakes, and woodland and estate walks.

OPEN: 28 March to 2 Nov: daily except Tues & Sat, weekdays 12–5.30; Sun & BH Mon 11–5.30. Last admission 5. Car park open dawn to dusk throughout the year. Estate walks leaflet available.

ADMISSION: £4, children £2, family ticket £10. Pre-booked parties of 20 or more weekday afternoons £3 (no reduction Sun & BH).

HAM HOUSE

THE NATIONAL TRUST

The National Trust was founded in 1895 to prot the best of our heritage for ever. It is not a government department, but an independent charity with a membership open to all. It relies thousands of volunteers who help in every aspe the Trust's work. Every visit you make to a National Trust property is a donation to the wo of the charity.

Today the Trust protects 350 houses and gardens over 550 miles of coastline and more than 238,0 hectares (600,000 acres) of countryside - all preserved for present and future generations to enjoy.

This map guide features just some of the Natior Trust's finest houses and gardens in the South E of England and provides details of opening time admission prices and location. Many of these properties have restaurants or tea rooms and sho which are indicated by the following symbols

Further details about these and many other properties across England, Wales and Northern Ireland can be found in the National Trust handbook which is available from all Trust properties and shops as well as high street bookshops. Alternatively you can contact one c the addresses on the back of this leaflet to find c more about specific properties in the South Eas England.

This montage shows the National Trust place can visit in and around London. For the admission charges and times please contac National Trust direct.

CARLYLE'S HOUSE

24 Cheyne Row, Chelsea SW3 5HL
Tel: 0171 352 7087

LOCATION: Off Cheyne Walk, between Battersea and Albert B on Chelsea Embankment, or off the King's Road and Oakley Stre *Station:* Victoria 1ml *Underground:* Sloane Sq 1ml.

The Chelsea hon Victorian writer historian, Thoma Carlyle, and his w Jane. Original furnishings, book pictures and pers items fill the inti interiors and vivi depict the lives o Carlyles. Dickens Chopin, Tennyso George Eliot and Emerson all cam The small, Victor garden which Ca enjoyed is also o delight today's vi

OPEN: 29 March 2 Nov: Wed to Su BH Mon 11–5. L admission 4.30. C Good Fri.

ADMISSION: £3 £1.50.

POLESDEN LACEY

HARTWELL

SILVER FURNITURE AT KNOLE

OSTERLEY PARK

CHARTWELL

Vesterham TN16 1PS
nformation Tel: 01732 866368

OCATION: 2ml S of Westerham, fork left off B2026 after 1½ml.
Station: Edenbridge 4ml; Edenbridge Town 4½ml; Oxted 5½ml;
evenoaks 6½ml.

Visit the family home where Britain's wartime Prime Minister, Sir Vinston Churchill, lived for more than 40 years. See rooms as they vere in Churchill's time, right down to the daily papers and his amous cigars. Capture the mood of key moments in 20th-century istory through photographs and books spanning Churchill's olourful career.

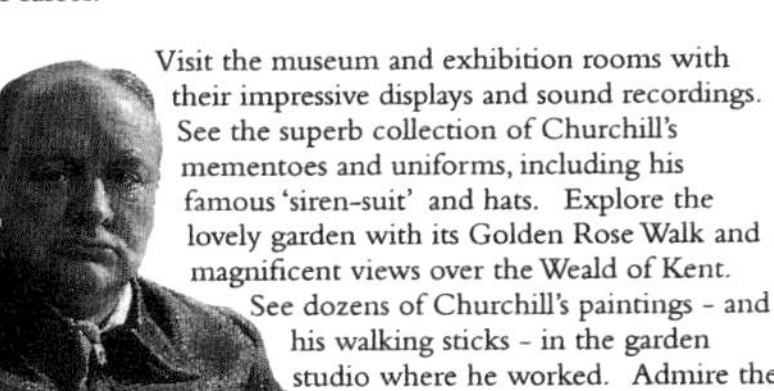

Visit the museum and exhibition rooms with their impressive displays and sound recordings. See the superb collection of Churchill's mementoes and uniforms, including his famous 'siren-suit' and hats. Explore the lovely garden with its Golden Rose Walk and magnificent views over the Weald of Kent. See dozens of Churchill's paintings - and his walking sticks - in the garden studio where he worked. Admire the garden walls that Churchill built with is own hands and the pond where he sat to feed the Golden Orfe. Gift shop, restaurant and function room.

)PEN: House & garden: March & Nov: Sat, Sun & Wed 11–4.30. ast admission 4. **House, garden & studio**: 29 March to 2 Nov: daily except Mon & Tues) 11–5.30. Last admission 4.30. Open BH Mon 1–5.30, last admission 4.30.

ADMISSION: House, garden & studio £5, children £2.50, family icket £12.50. Garden & studio only £2.50. March & Nov, house only £3. Coaches and groups by appointment only; no reductions.

West Clandon, Guildford
GU4 7RQ
Tel: 01483 222482

LOCATION: At West Clandon on A247, 3ml E of Guildford; if using A3 follow signposts to Ripley to join A247 via B221. *Station:* Clandon 1ml. Turn left on main road.

Clandon offers something for everyone with 14 magnificent showrooms, including the impressive marble hall, the famous Gubbay collection of porcelain, furniture and needlework and the Ivo Forde collection of Meissen Italian Comedy Figures. There is also the fascinating Museum of the Queen's Royal Surrey Regiment and, in the garden, a genuine Maori Meeting House. All this plus Clandon's celebrated restaurant, beautiful gardens and grounds for picnics.

OPEN: House (new times): 30 March to 30 Oct: Tues, Wed, Thur & Sun, plus BH Mon 11.30–4.30; last admission 4. **Garden**: daily 9–7.30 (or dusk if earlier). **Museum**: 30 March to 30 Oct: Tues, Wed, Thur, Sun & BH Mon 12–5.

ADMISSION: House & garden £4; family ticket £10. Combined ticket with Hatchlands Park £6. Groups welcome, Please book in advance.

CLAREMONT LANDSCAPE GARDEN

Portsmouth Road, Esher KT10 9JG
Tel: 01372 469421

LOCATION: On S edge of Esher, on E side of A307 (no access from Esher bypass). *Station:* Esher 2ml; Hersham 2ml; Claygate 2ml.

Open throughout the year, Claremont has something to delight you in every season. A stroll around the lake and island will take in the grotto, avenues, glades and the extraordinary turf amphitheatre. There are circular walks with wonderful views suitable for pushchairs and wheelchairs. Bring a picnic or sample the delights of Claremont's tea-room. Relax in this 18th-century landscape garden.

CLIVEDEN

KNOLE

Sevenoaks TN15 0RP
Infoline: 01732 450608

LOCATION: Off M25 London Orbital at S end of Sevenoaks town; just E of A225. *Station:* Sevenoaks 1½ml.

KNOLE

Tour 13 magnificent state rooms in one of the great treasure houses of England - a "calendar" house dating from the 15th century with 365 rooms, 52 staircases and seven courtyards (for days of the year, weeks of the year, and days of the week). Find the treasures of kings and queens: exquisite silver furniture, fragile tapestries, rare carpets, and other unique furniture including the first "Knole" settee, state beds and even an early royal loo! Study a remarkable collection of historical portraits by masters such as Sir Joshua Reynolds, Van Dyck and Gainsborough. Enjoy displays in the Brewhouse Tea Room.

Uncover Knole's intriguing history of connections with the famous: Archbishops of Canterbury, Henry VIII, Elizabeth I, the Sackville family, Dukes of Dorset, Vita Sackville-West and Virginia Woolf whose novel Orlando is set at the house. Enjoy throughout the year the extensive deer park owned by Lord Sackville.

OPEN: House: 28 March to 2 Nov: Wed, Fri, Sat, Sun & BH Mon 11–5; Thur 2–5. Last admission 4. Pre-booked groups accepted on Wed, Fri & Sat 11–4, Thur 2–4. **Park**: open daily to pedestrians by courtesy of Lord Sackville. **Garden**: May to Sept: first Wed in each month only, by courtesy of Lord Sackville, 11–4; last admission 3.

ADMISSION: House: £5; children £2.50; family ticket £12.50. Pre-booked parties £4. **Garden (note limited opening times)**: £1, children 50p.

SPECIAL OFFER: London Charing Cross to Knole combined train bus/admission ticket - up to five services daily each way, journey time less than one hour. Adult £10, child £5, concessions for NT members, families and Railcard holders. For train times and fares telephone 0345 484950

MORDEN HALL PARK

Morden Hall Road, Morden SM4 5JD
Tel: 0181 648 1845

LOCATION: Off A24, and A297 S of Wimbledon, N of Sutton. *Station:* Morden Road, not Sun, ½ml. *Underground:* Morden ½ml

A green oasis in the heart of SW London, this former deer park offers you the opportunity to get away from it all. Step back in time and enjoy the waterways, trees, Elizabethan meadow and the rose garden. The old estate workshops are again buzzing with activity with craft workshops to visit. Licensed riverside cafe serving coffee, lunches and teas daily. Independently managed City Farm. Garden Centre (not NT).

MORDEN HALL PARK

OPEN: Park only open all year during daylight hours. *Note:* Car park by the café/shop and garden centre closes at 6 daily.

ADMISSION: Free.

OSTERLEY PARK

Isleworth, Middlesex TW7 4RB
Tel: 0181 560 3918

LOCATION: Access via Thornbury Road on N side of A4 between Gillette Corner and Osterley underground station; M4, Jn 3. *Station:* Syon Lane 1½ml *Underground:* Osterley ¾ml.

Hidden away beside the A4 on the way to Heathrow Airport lies a unique country estate with an 18th-century neo-classical house by Robert Adam set in 140 ha (300 acres) of park and farmland. Osterley contains something for everyone, from the connoisseur of art and architecture to families, walkers and those who are simply looking to refresh the spirit. Stables Tea Room for light lunches and teas open from 11.30.

OPEN: House: 26 March to 2 Nov: Wed to Sun 1–5; BH Mon 11–5. Closed Good Fri. Last admission 4.30. **Grand Stable**: Sun afternoons in summer. **Park and pleasure grounds**: all year 9–7.30 or sunset if earlier. Park will close early during major events. Car park closed 25 & 26 Dec.

ADMISSION: £3.80; family ticket £9.50. Parties Wed to Sat £3, advance booking required. Park and pleasure grounds free. £1 off adult ticket for holders of valid LT travelcard. Car park £2, refundable on purchase of ticket to house, but free 3 Nov to March 1998.

PETWORTH HOUSE AND PARK

Petworth GU28 0AE
Tel: 01798 342207 Infoline: 01798 343929

LOCATION: In centre of Petworth (A272/A283); *Station:* Pulborough 5¼ml.

Perfect for a great day out, the magnificent late 17th-century house, set in 700 acres of beautiful deer park and pleasure grounds landscaped by 'Capability' Brown. The house contains the Trust's finest collection of paintings including works by Turner, Van Dyck, Reynolds and Blake, ancient and neo-classical sculpture, fine furniture and elaborate carvings by Grinling Gibbons. Old kitchens and service room recently opened in the servants' block. Chapel closed in 1997 for major restoration work. Coffee, light lunches and teas in licensed tea room.

OPEN: House: 28 March to 2 Nov: daily except Thur & Fri (but open Good Fri) 1–5.30. Last admission to house 4.30; old kitchens 5. Additional rooms shown weekdays (not BH Mon). **Pleasure grounds and car park** 12–6 (opens 11 on BH Mon and all of July & Aug) for walks, picnics and access to tea-room, shop and Petworth town. **Park:** open daily 8- dusk. No charge. Closed 27/28/29 June from 12 for open air concerts.

ADMISSION: £4.50; family ticket £12. Coach parties must book in advance; please contact Administration Office.

THE NATIONAL TRUST
PO Box 39, Bromley, Kent, BR1 3XL. Tel: 0181 315 1111

The National Trust is an independent registered charity (no. 205846)
All photographs from National Trust Photolibrary
Printed in England by Watmoughs Ltd

Explore England's History

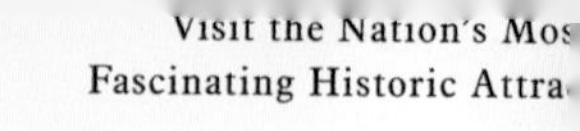

CHISWICK HOUSE

Recognised as one of the most glorious examples of an English building inspired by the architecture of ancient Rome.

Chiswick House contains fine interiors and boasts spectacular 18th century Italianate gardens with classical statues and neoclassical temples.

Chiswick: Great West Road (A4), London.
Adults £3.00, concessions £2.30, children £1.50.
☎ **0181 995 0508.**

London & South East

Explore the magnificent historic houses of the country's capital containing many fine works of art, as well as a wealth of historic sites in the surrounding counties.

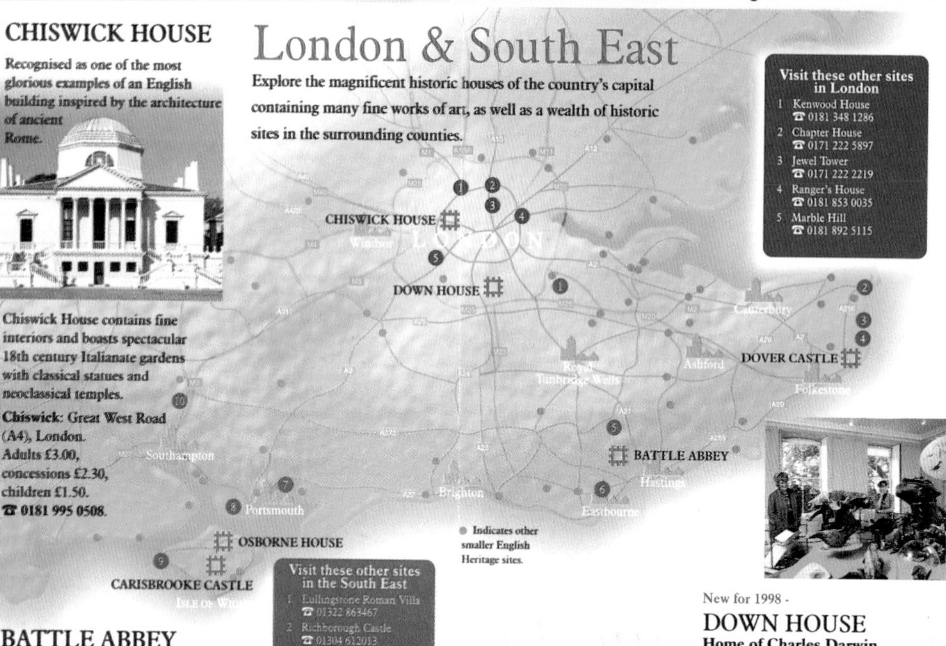

Visit these other sites in London

1. Kenwood House ☎ 0181 348 1286
2. Chapter House ☎ 0171 222 5897
3. Jewel Tower ☎ 0171 222 2219
4. Ranger's House ☎ 0181 853 0035
5. Marble Hill ☎ 0181 892 5115

Visit these other sites in the South East

1. Lullingstone Roman Villa ☎ 01322 863467
2. Richborough Castle ☎ 01304 612013
3. Deal Castle ☎ 01304 372762
4. Walmer Castle & Gardens ☎ 01304 364288
5. Bayham Abbey ☎ 01892 890381
6. Pevensey Castle ☎ 01323 762604
7. Portchester Castle ☎ 01705 378291
8. Fort Brockhurst ☎ 01705 581059
9. Yarmouth Castle ☎ 01983 760678
10. Wolvesey Castle ☎ 01962 854766

BATTLE ABBEY

Discover the story of the last successful invasion of England and the most famous date in English history - 1066 and the Battle of Hastings. Experience the battle at its fiercest with our interactive audio tour.

Battle Abbey: situated at the end of the High Street, Battle. Adults £4.00, concessions £3.00, children £2.00.
☎ **01424 773792.**

New for 1998 -

DOWN HOUSE

Home of Charles Darwin.

Discover the fascinating life and works of one of the greatest scientific thinkers through a variety of intriguing presentations, exhibits and interactive displays in this carefully refurbished home.

Down House: In Luxted Road, Downe off A21 near Biggin Hill.
Adults £5.00, concessions £3.80, children £2.50.
☎ **01689 859119.**

Closed Mondays,Tuesdays and February. House by guided tour only, booking essential.
Booking Line: 0870 603 0145

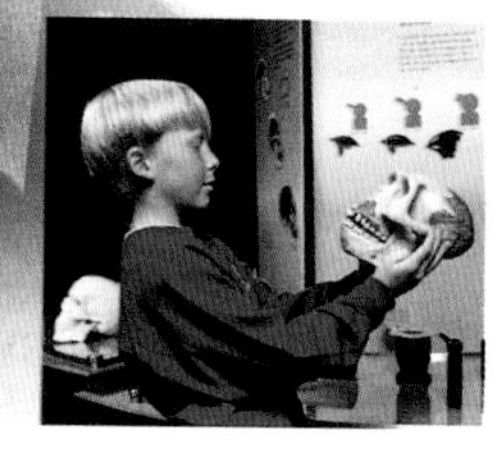

OSBORNE HOUSE Isle of Wight

Discover the favourite home of Queen Victoria, Britain's longest reigning monarch. This magnificent house in extensive grounds remains very much as it was when Victoria died here in 1901. A truly revealing insight into royal family life.

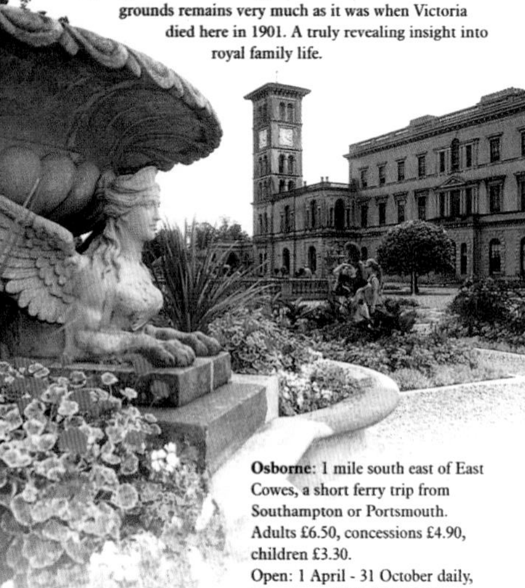

Osborne: 1 mile south east of East Cowes, a short ferry trip from Southampton or Portsmouth.
Adults £6.50, concessions £4.90, children £3.30.
Open: 1 April - 31 October daily, 10am-5pm. ☎ **01983 200022.**

CARISBROOKE CASTLE Isle of Wight

Enjoy a day of discovery at the former prison of King Charles I who was held here before his execution in 1648. Admire breathtaking views, see the beautiful chapel, a fascinating museum and the famous Carisbrooke donkeys.

Carisbrooke: 1¼ miles south west of Newport, a short ferry trip from Southampton or Portsmouth.
Adults £4.00, concessions £3.00, children £2.00.
☎ **01705 378291**

DOVER CASTLE, SECRET WARTIME TUNNELS AND SIEGE OF 1216

Delve into 2,000 years of colourful history at one of the nation's most exciting heritage attractions. Experience wartime drama deep in the heart of the White Cliffs of Dover, where top secret tunnels played a vital role in military action. Commentary, film, sound and smell will enable you to re-live those remarkable days of World War II.

New for 1998 - **Experience a Castle under a Medieval Siege**
A thrilling new computer programmed sound and light presentation combined with three dimensional displays evoke the terrifying reality of this momentous siege.

Dover Castle: The castle is situated just above the port of Dover.
Adults £6.60, concessions £5.00, children £3.30.
☎ **01304 201628.**

The North

The rugged and undulating beauty of the North counties contain much of the country's finest h
From Hadrian's Wall, the greatest testimony to Roman occupation, to the enchanting priory of Lindisfarne on Holy Island.

HADRIAN'S WALL

The greatest monument to the history of the Roman occupation stretching across northern England from the Solway Firth to the River Tyne. Designated a UNESCO World Heritage Site, you can explore three of the best preserved settlements at Chesters, Corbridge and Housesteads.

Admi
us, tele
Chesters Roman
Corbridge Roman
Housesteads Roman

The Midlands

Discover the very Heart of England and a regio abbeys and elegant stately homes with their glo gardens.

STOK CAST

A perfectly century m impressive medieval b

Stokesay: of Ludlow. concession
☎ **01588**

KENILWORTH CASTLE

England's finest and most extensive castle ruins. Th names and events of history enhance its rich and il

K
ir
A
cc
cl
☎

The South West

The lovely counties of Cornwall, Devon, Wiltsh Somerset hold many historic secrets, most notab legends of King Arthur and his Round Table at prehistoric mysteries of Stonehenge.

STONEHENGE

The great and ancient stone circle of Stonehenge is one of the wonders of the world and as old as many of the temples and pyramids of Egypt. An impressive and enigmatic prehistoric monument and a treasured World Heritage Site of unique importance.

Stonehenge: 2 miles west of Amesbury. Adults £3.9 concessions £2.90, children £2.00. ☎ **01980 623 108**

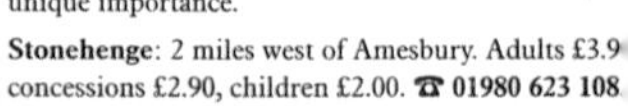

DISCOVER OVER 400 ENGLISH HERITAGE ATTRACTIONS

LINDISFARNE PRIORY

One of the holiest of Anglo-Saxon sites and the 'Cradle of Christianity' for the nation.
The beautiful Priory and its fascinating museum are situated on Holy Island, only accessible by causeway at low tide.

Lindisfarne: on Holy Island. Adults £2.70, concessions £2.00, children £1.40.
☎ 01289 389200.

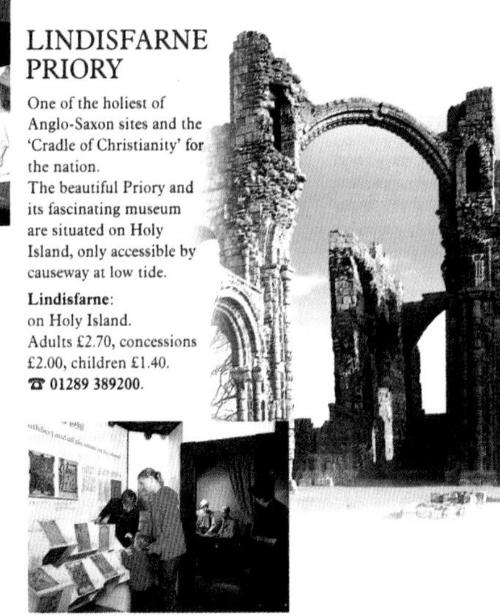

BELSAY HALL, CASTLE & GARDENS

A neo-classical hall and 14th century castle set in 30 acres of magnificently landscaped grounds. The spectacular Quarry Garden is breathtaking and contains a glorious array of rare plant life.

Belsay Hall: in Belsay. Adults £3.60, concessions £2.70, children £1.80.
☎ 01661 881636.

Visit these other sites in the North

1. Dunstanburgh Castle ☎ 01665 576231
2. Warkworth Castle ☎ 01665 711423
3. Carlisle Castle ☎ 01228 591922
4. Richmond Castle ☎ 01748 822493
5. Mount Grace Priory ☎ 01609 883494
6. Whitby Abbey ☎ 01947 603568
7. Scarborough Castle ☎ 01723 372451
8. Pickering Castle ☎ 01751 474989
9. Brodsworth Hall ☎ 01302 722598
10. Beeston Castle ☎ 01829 260464

Visit these other sites in the Midlands

1. Peveril Castle ☎ 01433 620613
2. Bolsover Castle ☎ 01246 823349
3. Hardwick Old Hall ☎ 01246 850431
4. Castle Rising Castle ☎ 01553 631330
5. Wroxeter Roman City ☎ 01743 761330
6. Kirby Hall ☎ 01536 203230
7. Witley Court ☎ 01299 896636
8. Wrest Park & Gardens ☎ 01525 860152
9. Orford Castle ☎ 01394 450472
10. Goodrich Castle ☎ 01600 890538

Visit these other sites in the South West

1. Old Wardour Castle ☎ 01747 870487
2. Sherborne Old Castle ☎ 01935 812730
3. Portland Castle ☎ 01305 820539
4. Berry Pomeroy Castle ☎ 01803 866618
5. Totnes Castle ☎ 01803 864406
6. Dartmouth Castle ☎ 01803 833588
7. Launceston Castle ☎ 01566 772 365
8. Restormel Castle ☎ 01208 872687
9. St. Mawes Castle ☎ 01326 270526
10. Chysauster Ancient Village ☎ 0831 757 934

● Indicates other smaller English Heritage sites. For full details see our *Visitors' Handbook* available from all staffed sites.

RIEVAULX ABBEY

Spectacular and extensive remains of the first Cistercian monastery in northern England set in the beautiful and serene valley of the River Rye.

Rievaulx Abbey: in Rievaulx, 2¼ miles west of Helmsley. Adults £2.90, concessions £2.20, children £1.50.
☎ 01439 798228.

CLIFFORD'S TOWER

A proud testament to York's medieval castle with fine views of the historic city and sheltering a shameful and bloody past.

Clifford's Tower: in the centre of York opposite Castle Museum. Adults £1.70, concessions £1.30, children 90p.
☎ 01904 646940.

AUDLEY END HOUSE & GARDENS

A palatial Jacobean mansion in a glorious parkland setting featuring 30 exquisitely furnished rooms.

Audley End: 1 mile west of Saffron Walden. Adults £5.95, concessions £4.50, children £3.00. Open: 1 April - 30 September 11am-6pm. 1-31 October, 10am-3pm. Closed Mondays & Tuesdays. Open Bank Holidays. **☎ 01799 522 399**.

FRAMLINGHAM CASTLE

An impressive 12th century castle with superb views of the countryside. Formerly used as a military fortress, an Elizabethan prison, a poor house and as a school.

Framlingham: in Framlingham. Adults £2.95, concessions £2.20, children £1.50. **☎ 01728 724189**.

Bringing History Alive!

With the Overseas Visitor Pass it's so much easier to experience the excitement of an English Heritage event. From April to October history comes alive as we stage over 600 top quality events including living history, battle re-enactments, music and drama.

To find out more call:

☎ +44 (0)171 973 3434

PENDENNIS CASTLE

Enjoying glorious views over the mile wide River Fal, the castle has stood in defence of these shores for almost 450 years.

New for 1998 - The complete 450 year history revealed for the very first time. Return to the 15th century in the mighty Keep and see a Tudor gun-deck in action. Explore the hands-on Discovery Centre and fully restored World War II operations centre and experience garrison life at the turn of the century complete with sounds and smells.

Pendennis: on Pendennis Head 1 mile south east of Falmouth. Adults £3.00, concessions £2.30, children £1.50.
☎ 01326 316594.

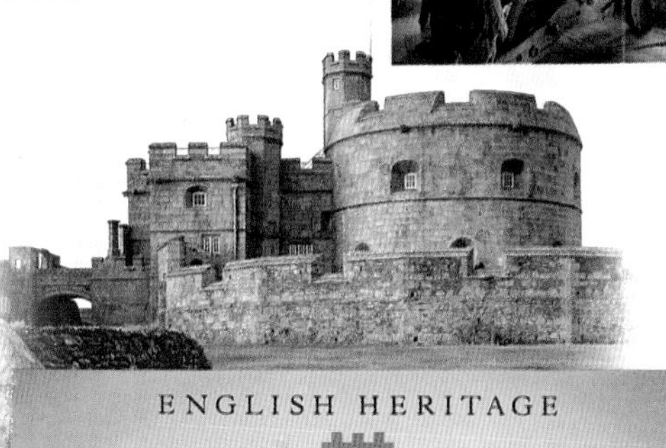

OLD SARUM

Founded around 500BC this once vast settlement was occupied through the centuries by Romans, Saxons and Normans and today evokes a powerful atmosphere.

Old Sarum: 2 miles north of Salisbury. Adults £2.00, concessions £1.50, children £1.00.
☎ 01722 335398.

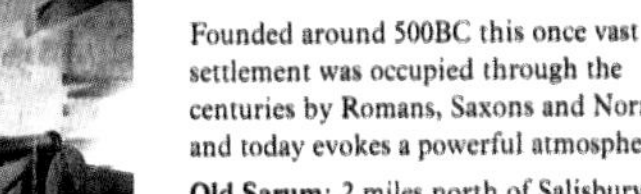

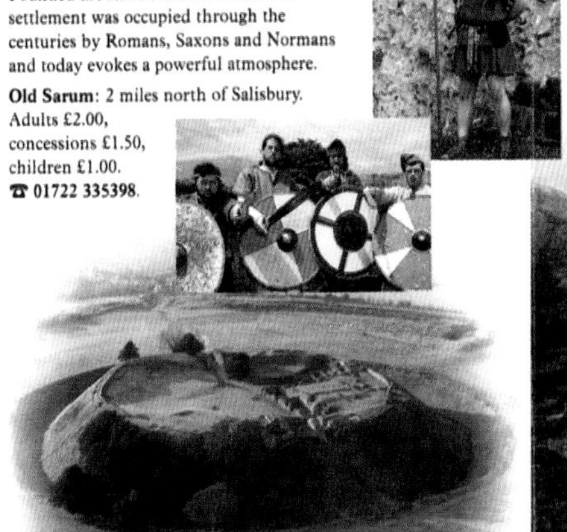

TINTAGEL CASTLE

A remarkable place of dreams, romance and myths set in one of the country's most spectacular coastal locations and rich with the legends of King Arthur and the magician Merlin.

Tintagel: from Tintagel Village, walk along path to Tintagel Head. Adults £2.80, concessions £2.10, children £1.40.
☎ 01840 770328.

Acknowledgements

For the writers, photographers, architects, engineers, artists, press officers and organisers of the millennium exhibition who have supplied rich sources of information, I acknowledge my deep gratitude during the research work. The credit for the computer images of the dome exhibitions is due to NMEC/SHAM and NMEC/DAVIDSON.

I would like to express my thanks to my institution, the University of East London, for its support of the research work over a number of years. I am deeply indebted to numerous individuals, previous writers, other photographers, estate agents and many diverse organisations who so kindly helped with the preparation of this book.

For the supply of considerable information and illustrations, I am deeply grateful to the New Millennium Experience Company, NMEC Press Office, Jubilee Line/QA photos, the former London Docklands Development Corporation (LDDC) and the Port of London Authority (PLA). To Canary Wharf Development Company and Olympia & York special thanks are given for the supply of many photographs, maps and illustrations.

For the generous supply of superb aerial photographs I am most grateful to Chorley and Handford, Stafford Road, Wallington, Surrey. The assistance and co-operation of Tom Samson and Paul Proctor are most appreciated.

Acknowledgement with thanks has to be made for information received from the Museum of London and the Museum of Docklands, Thames Water, National Trust and English Heritage.

Much appreciation is due to the staff of the Guildhall Library and the Public Relations Department of the City of London Corporation for the supply of information and slides.

For general assistance I am grateful to Terence O'Connell. Special thanks are due to Tom Juffs for his enthusiasm and unstinting support throughout the project. I would like to thank sincerely Linda Day for her help, excellent typing and patience in preparation of the whole manuscript with great care. I would like to express my deep gratitude to Dr John Grubert for proof reading and assistance with the layout of illustrations in the book. I am grateful to Dr Paul Smith for advice and to Joyce O'Neill for continued help. Thanks are due to Dave Hobson and his staff of Lipscomb Printers. I am most grateful to my wife, Irene, for her continued support and lasting patience over many years.

Considerable information was supplied by many organisations including Docklands News, Docklands Digest, Greenwich Magazine, Port of London Magazine, Esso Magazine, Rob Poulton, New Civil Engineer, The British Library, British Museum, University of Edinburgh Library, John Tramper, Jurassic Park, Natural History Museum, Science Museum, BBC, Paul Wade, Readers Digest, Butlers Wharf Limited, St Martins Property Corporation, St Katharine by the Tower, Jacobs Island Company, The Anchor Brewhouse, Barratt East London, Costain Homes, Heron Homes, Laing Homes, London Transport Museum, Waterside Inn of Luton, Waites Homes, Wimpey Homes, Rosehaugh co-partnership, Trafalgar House Residential, Colin Druce & Company, Docklands Light Railway, John Mowlem, Thames Line, the Daily Telegraph, Daily Mirror, the Independent and Times newspapers, Evans Evans Tours, Original London Sightseeing Tour, Golden Tours, Tower of London, English Tourist Board, London Tourist Board, Greenwich Council and Tourist Board, Millennium Exhibition Trust, Millennium Commission, National Curriculum Council, Hanover Expo 2000, Newmarket Journal, Thames Line, Savilles Property, Prudential Property, Carleton Smith, Clapshaws, Cluttons, Shakespeare Globe Theatre, National Maritime Museum, National Rivers Authority, Trafalgar House and Hayes Davidson for the dome computer image.

To members of the public, visitors, students, teachers and scholars world-wide, who have kindly supported our book publications over the past two decades, some of which are in their seventh and eighth editions, I express my deepest appreciation. The books are providing an essential public service.

Information

A unique set of seven internationally acknowledged research books have been published on the history, heritage, regeneration, infrastructure and millennium of London and Docklands. They are ideal for teaching and research in schools and colleges as well as for libraries, visitors and the public.

"Dockland"Historical Survey,
'ISBN 0-901987-8)
'London Docklands" Past, Present and Future (ISBN 0-091987-81-6),
'Discover London Docklands" A-Z Illustrated Guide to Modern Docklands (ISBN 1-874536-00-7),
'European Docklands", Past, Present and Future (ISBN 0-901987-82-4),
London Illustrated" Historical, Current and Future.
(ISBN 1-874536-01-5),
'London Docklands Guide" Heritage and Millennium Exhibition (ISBN 1-874536-03-1).
"London Millennium Guide" Education, Entertainment and Aspirations.
(ISBN 1-8745-36-20-1).

For further information visit the University Web Site on www.uel.ac.uk or telephone 0181 849 3580, Fax 0181 849 3423..

PLEASE ORDER THROUGH:
RESEARCH BOOKS,
P O. BOX 82,
ROMFORD, ESSEX,
RM8 2AS, GREAT BRITAIN

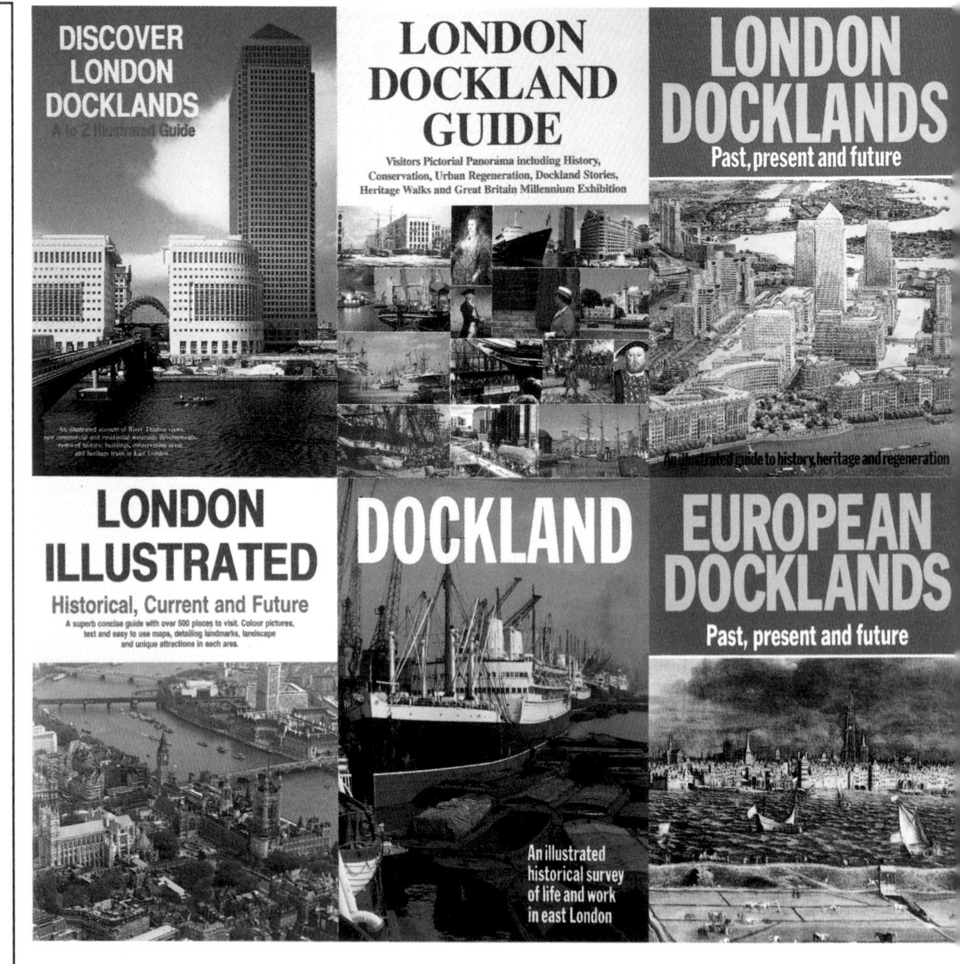

London and Docklands Research Books
The Essential Collection for Londoners and Visitors